Fossils At A Glance

This book is dedicated to Brian Rigby, 1937–2002

Fossils At A Glance

Clare Milsom

Liverpool John Moores University

Sue Rigby

University of Edinburgh

350 Main Street, Malden, MA 02148-5020, USA
108 Cowley Road, Oxford OX4 1JF, UK
550 Swanston Street, Carlton,
Victoria 3053, Australia

First published 2004 by Blackwell Science Ltd, a Blackwell Publishing company

Library of Congress Cataloging-in-Publication Data

Milsom, Clare.
Fossils at a glance/Clare Milsom, Sue Rigby.
p. cm.
Includes index.
ISBN 0-632-06047-6 (pbk. : alk. paper)
1. Paleontology. 2. Fossils. I. Rigby, Susan. II. Title.
QE711.3 .M55 2004
560–dc21 2002153751

A catalogue record for this title is available from the British Library.

Set in 9/12pt Galliard
by Graphicraft Limited, Hong Kong
Printed and bound in the United Kingdom
by TJ International Ltd, Padstow, Cornwall

For further information on
Blackwell Publishing, visit our website:
http://www.blackwellpublishing.com

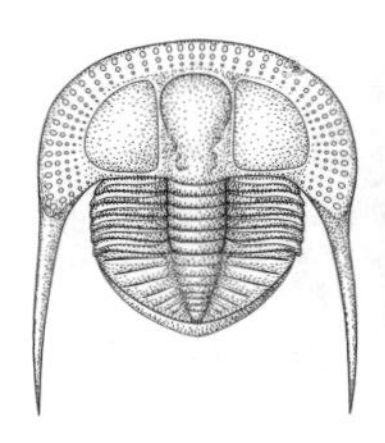

Contents

Acknowledgments, vi

Chapter 1 Introduction, 1
Chapter 2 Fossil classification and evolution, 11
Chapter 3 Sponges, 17
Chapter 4 Corals, 21
Chapter 5 Bryozoans, 31
Chapter 6 Brachiopods, 35
Chapter 7 Echinoderms, 43
Chapter 8 Trilobites, 53
Chapter 9 Molluscs, 61
Chapter 10 Graptolites, 79
Chapter 11 Vertebrates, 87
Chapter 12 Land plants, 101
Chapter 13 Microfossils, 113
Chapter 14 Trace fossils, 127
Chapter 15 Precambrian life, 131
Chapter 16 Phanerozoic life, 137

Reading list, 147
Geological timescale, 148
Index, 149

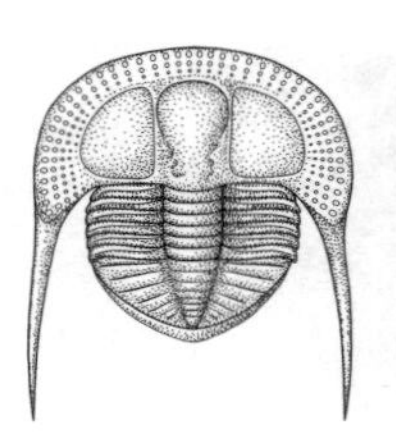

Acknowledgments

We are grateful to the following people for reading parts of the manuscript and improving it significantly: Chris Settle, Chris Paul, Liz Hide, Paul Taylor, Robin Cocks, Graham Budd, Liz Harper, Ivan Sansom, Jason Hilton, Bridget Wade, Simon Braddy, Nick Butterfield, and Sarah Gabbott.

In addition, our families provided endless distractions and generous support during the writing of this book. Thanks are due in this respect to Maurizio Bartozzi and to Michael, Peter, and Tom Fuller.

Figure acknowledgements

Figure 1.1: from *Treatise on Invertebrate Paleontology*, part O, Geol. Soc. Amer. and Univ. Kansas Press (Figure 159.6, O218); Crimes, T.P., Legg, I., Marcos, A. and Arboleya, M., 1977, in Crimes, T.P. and Harper, J.C. (eds) *Trace Fossils 2*, Seel House Press, Liverpool (Figure 10, p. 134). Figure 1.2: based on Williams, S.H., 1986, in Hughes, C.P. and Rickards, R.B., *Palaeoecology and Biostratigraphy of Graptolites*, Geological Society Special Publication 20 (Figure 1, pp. 166–7) and Barnes, C.R. and Williams, S.H., 1990, in Briggs, D.E.G. and Crowther, P.R. (eds) *Palaeobiology: A Synthesis*, Blackwell Scientific Publications (Figure 1, p. 479). Figure 1.3: based on various sources. Figure 1.4: modified and redrawn from Campbell, N.A., 1996, *Biology*, 4th edn (Figure 23-10, p. 468). Figure 1.5: modified and redrawn from Brenchley, P.J. and Harper, D.A.T., 1998, *Palaeoecology: Ecosystems, Environment and Evolution* (Figure 6.5, p. 184). Figure 1.6: modified and redrawn from Anderton, R., Bridges, P.H., Leeder, M.R. and Sellwood, B.W., 1993, *A Dynamic Stratigraphy of the British Isles: A Study in Crustal Evolution* (Figure 3.5, p. 34) and Williams, G.E., 1969, *Journal of Geology*, 77, 183–207. Figure 1.9: redrawn and modified from Seilacher, A., Reif, W.-E. and Westphal, F., 1985, *Philosophical Transactions of the Royal Society of London*, B11, 5–23. Figure 1.10: redrawn from *Treatise on Invertebrate Paleontology*, part A, Geol. Soc. Amer. and Univ. Kansas Press (Figure 7, A13). Figure 1.11: redrawn from Milsom, C.V. and Sharp, T., 1995, *Geology Today*, 11, 22–6.

Figure 2.2: redrawn from *British Mesozoic Fossils*, British Musuem (Natural History) (Plate 28(2)). Figure 2.3: modified and redrawn from Campbell, N.A., 1996, *Biology*, 4th edn (Figure 23-15, p. 476). Figure 2.4: modified and redrawn from Skeleton, P., 1993, *Evolution: A Biological and Palaeontological Approach*, Addison Wesley (Figure 11.1, p. 512).

Figure 3.1a, b: redrawn from Benton, M. and Harper, D., 1997, *Basic Palaeontology*, Addison Wesley Longman (Figure 5.10); Figure 3.1c, d: redrawn and simplified from

McKinney, F.K., 1991, *Exercises in Invertebrate Paleontology*, Blackwell Scientific Publications (Figures 4.1, 4.2). Figure 3.2: redrawn and simplified from Prothero, D.R., 1998, *Bringing Fossils to Life*, W.C.B./McGraw-Hill USA (Figure 12.7). Figure 3.3: redrawn and simplified from Clarkson, E.N.K., 1998, *Invertebrate Palaeontology and Evolution*, Chapman and Hall, London (Figure 4.16a). *Siphonia*, *Rhaphidonema*: from Clarkson, E.N.K., 1998, *Invertebrate Palaeontology and Evolution*, Chapman and Hall, London (Figure 4.6a, d).

Figure 4.2: simplified from Clarkson, E.N.K., 1998, *Invertebrate Palaeontology and Evolution*, Chapman and Hall, London (Figure 5.20). Figure 4.3: redrawn from McKinney, F.K., 1991, *Exercises in Invertebrate Paleontology*, Blackwell Scientific Publications (Figure 5.2d). Figure 4.4a: redrawn from McKinney, F.K., 1991, *Exercises in Invertebrate Paleontology*, Blackwell Scientific Publications (Figure 5.6); Figure 4.4b: redrawn from various sources. Figure 4.5: after McKinney, F.K., 1991, *Exercises in Invertebrate Paleontology*, Blackwell Scientific Publications (Figures 5.4, 5.6). Figure 4.6: redrawn and simplified from Prothero, D.R., 1998, *Bringing Fossils to Life*, W.C.B./McGraw-Hill USA (Figure 12.13). Figure 4.7: redrawn from *British Palaeozoic Fossils*, British Museum (Natural History) (Plate 3(7)). Figure 4.8: simplified from Clarkson, E.N.K., 1998, *Invertebrate Palaeontology and Evolution*, Chapman and Hall, London (Figure 5.7f). Figure 4.9: based on various sources. *Favosites*, *Halysites*: redrawn from *British Palaeozoic Fossils*, British Museum (Natural History) (Plate 15(1,3)). *Palaeosmilia*: redrawn from *British Palaeozoic Fossils*, British Museum (Natural History) (Plate 44(6)). *Isastraea*, *Montlivaltia*, *Thecosmilia*: redrawn from *British Mesozoic Fossils*, British Museum (Natural History) (Plate 3(2,4,6)). *Lithostrotion*, *Dibunophyllum*: redrawn from *British Palaeozoic Fossils*, British Museum (Natural History) (Plate 43(1,2)).

Figure 5.1a, c: simplified and redrawn from Benton, M. and Harper, D., 1997, *Basic Palaeontology*, Addison Wesley Longman (Figure 6.34); Figure 5.1b: redrawn from McKinney, F.K., 1991, *Exercises in Invertebrate Paleontology*, Blackwell Scientific Publications (Figure 12.3). Figure 5.2: simplified from Clarkson, E.N.K., 1998, *Invertebrate Palaeontology and Evolution*, Chapman and Hall, London (Figure 6.7). *Fenestella*: redrawn from *British Palaeozoic Fossils*, British Museum (Natural History) (Plate 4(1)). *Stomatopora*: redrawn from Clarkson, E.N.K., 1998, *Invertebrate Palaeontology and Evolution*, Chapman and Hall, London (Figure 6.11d).

Figure 6.1a: redrawn and modified from Prothero, D.R., 1998, *Bringing Fossils to Life*, W.C.B./McGraw-Hill, USA (Figure 13.2C, p. 228); Figure 6.1b: redrawn and modified from Clarkson, E.N.K., 1998, *Invertebrate Palaeontology and Evolution*, Chapman and Hall, London (Figure 7.1e, f, p. 159). Figure 6.2a: redrawn and modified from Clarkson, E.N.K., 1998, *Invertebrate Palaeontology and Evolution*, Chapman and Hall, London (Figure 7.1a, b, p. 159); Figure 6.2b: redrawn and modified from Clarkson, E.N.K., 1998, *Invertebrate Palaeontology and Evolution*, Chapman and Hall, London (Figure 7.5a, p. 165). Figure 6.3: based on Ziegler, A.M., Cocks, L.R.M. and Bambach, R.K., 1968, *Lethaia*, 1, 1–27. *Lingula*: redrawn from Black, R., 1979, *The Elements of Palaeontology*, Cambridge University Press (Figure 91a, p. 149). *Megellania*: redrawn from Clarkson, E.N.K., 1998, *Invertebrate Palaeontology and Evolution*, Chapman and Hall, London (Figure 7.1d, p. 159). *Gigantoproductus*: redrawn from *British Palaeozoic Fossils*, British Museum (Natural History) (Plate 47(6)). *Pentamerus*: redrawn from *British Palaeozoic Fossils*, British Museum (Natural History) (Plate 17(10)). *Spirifer*: redrawn from Black, R., 1979, *The Elements of Palaeontology*, Cambridge University Press (Figure 93a, d, p. 152). *Prorichthofenia*: redrawn from Black, R., 1979, *The Elements of Palaeontology*, Cambridge University Press (Figure 92j, p. 151). *Tetrarynchia*: redrawn from Black, R., 1979, *The Elements of Palaeontology*, Cambridge University Press (Figure 94a, b, p. 155). *Colaptomena*: redrawn from McKinney, F.K., 1991, *Exercises in Invertebrate Palaeontology*, Blackwell Scientific Publications (Figure 11.7, p. 160).

Figure 7.1: courtesy of H. Hess; Hess, H., Ausich, W.I., Brett, C.E. and Simms, M.J., 1999, *Fossil Crinoids*, Cambridge University Press (Figure 90, p. 78). Figure 7.2: redrawn and modified from Clarkson, E.N.K., 1998, *Invertebrate Palaeontology and Evolution*, Chapman and Hall, London (Figure 9.34b, p. 265). Figure 7.3: redrawn and modified from Moore, J., 2001, *Introduction to the Invertebrates*, Cambridge University Press (Figure 17.4ei, p. 271). Figure 7.4: redrawn from McKinney, F.K., 1991, *Exercises in Invertebrate Palaeontology*, Blackwell Scientific Publications (Figure 13.1, p. 186). Figure 7.5: based on McKinney, F.K., 1991, *Exercises in Invertebrate Palaeontology*, Blackwell Scientific Publications (Figure 13.6, p. 192). Figure 7.6: redrawn and modified from Moore, J., 2001, *Introduction to the Invertebrates*, Cambridge University Press (Figure 17.4eii, p. 271). Figure 7.7: based on various sources. All echinoderms redrawn from *British Palaeozoic Fossils*, British Museum (Natural History) (Plates 59(8,9); 60(3)); *British Mesozoic Fossils*, British Museum (Natural History) (Plates 42(1,2), 44(1), 45(2), 69(1), 70(3)).

Figure 8.1: based on various sources. Figure 8.3: based on various sources. *Paradoxides*, *Trinucleus*, *Agnostus*, *Dalmanites*, *Calymene*, *Cyclopyge*, *Phillipsia*, *Deiphon*: redrawn from *British Palaeozoic Fossils*, British Museum (Natural History).

Figure 9.1: redrawn and modified from Moore, J., 2001, *Introduction to the Invertebrates*, Cambridge University Press (Figure 10.1a, p. 132). Figure 9.2: based on various sources. Figure 9.3a, b: redrawn and modified from Prothero, D.R., 1998, *Bringing Fossils to Life*, W.C.B./McGraw-Hill, USA (Figure 15.11F, G, p. 288); Figure 9.3c: redrawn and modified from Moore, J., 2001, *Introduction to the Invertebrates*, Cambridge University Press (Figure 11.1d, p. 153). Figure 9.4: based on various sources. Figure 9.5: based on Clarkson, E.N.K., 1998, *Invertebrate Palaeontology and Evolution*, Chapman and Hall, London (Figure 8.11, p. 211). Figures 9.6, 9.7: redrawn and modified from Stanley, S.M., 1968, *Journal of Paleontology*, 42, 214–29 (Figures 6 and 4, respectively). Figure 9.10: redrawn and modified from Boss, K.J., 1982, in Parker, S.P. (ed.) *Synopsis and Classification of Living Organisms*, McGraw-Hill, New York (p. 968). Figure 9.8: redrawn and modified from Moore, J., 2001, *Introduction to the Invertebrates*, Cambridge University Press (Figure 11.4a, p. 160). Figure 9.9: redrawn from Doyle, P., 1996, *Understanding Fossils*, Wiley and Sons, UK (Figure 9.12, p. 172). Figure 9.10: redrawn and modified from Boss, K.J., 1982, in Parker, S.P. (ed.) *Synopsis and Classification of Living Organisms*, McGraw-Hill, New York (p. 1088). Figure 9.11: redrawn and modified from Clarkson, E.N.K., 1998, *Invertebrate Palaeontology and Evolution*, Chapman and Hall, London (Figure 8.21a, p. 231). Figure 9.12: redrawn and modified from Clarkson, E.N.K., 1998, *Invertebrate Palaeontology and Evolution*, Chapman and Hall, London (Figure 8.24c, p. 239). Figure 9.14: redrawn from Callomon, J.H., 1963, *Transactions of the Leicester Literary and Philosophical Society*, 57, 21–6. Figure 9.15: redrawn from Clarkson, E.N.K., 1998, *Invertebrate Palaeontology and Evolution*, Chapman and Hall, London (Figure 8.30, p. 29); *Treatise on Invertebrate Paleontology*, part L, Geol. Soc. Amer. and Univ. Kansas Press. Figure 9.16: redrawn and modified from Benton, M. and Harper, D. 1997, *Basic Palaeontology*, Addison Wesley Longman (Figure 8.31a, p. 188). Figure 9.17: redrawn and modified from Prothero, D.R., 1998, *Bringing Fossils to Life*, W.C.B./McGraw-Hill, USA (Figure 15.27, p. 304); Batt, R.J., 1989, *Palaios*, 4, 32–42. Figure 9.18: redrawn and modified from Brusca, R.C. and Brusca, G.J., 1990, *Invertebrates*, Sinauer Associates, USA (Figure 13G, p. 712). Figure 9.19: redrawn from Clarkson, E.N.K., 1998, *Invertebrate Palaeontology and Evolution*, Chapman and Hall, London (Figure 8.32a–c, p. 253). *Mya: British Caenozoic Fossils*, British Museum (Natural History) (Plate 38(11)). *Ensis*: redrawn from Clarkson, E.N.K., 1998, *Invertebrate Palaeontology and Evolution*, Chapman and Hall, London (Figure 8.11m, p. 211). *Teredo*: redrawn and modified from Black, R.M., 1970, *The Elements of Palaeontology*, Cambridge University Press (Figure 21b, p. 44). *Radiolites*: redrawn from Clarkson, E.N.K., 1998, *Invertebrate Palaeontology and Evolution*, Chapman and Hall, London (Figure 8.13j). *Turritella, Planorbis, Hygromia: British Caenozoic Fossils*, British Museum (Natural History) (Plates 39(5), 41(15), 42(2)). *Patella*: redrawn and modified from Black, R.M., 1970, *The Elements of Palaeontology*, Cambridge University Press

(Figure 35a, p. 64). *Ammonites*: redrawn from *British Mesozoic Fossils*, British Museum (Natural History) (Plates 30(2), 32(1,2), 37(4), 66(2)) and *British Palaeozoic Fossils*, British Museum (Natural History) (Plate 58(6)). *Neohibolites: British Mesozoic Fossils*, British Museum (Natural History) (Plate 67(4)).

Figure 10.1: based on various sources.

Figures 11.1, 11.2, 11.5, 11.7, 11.9, 11.10: original diagrams, with cartoons of skeletons redrawn from a variety of sources, most commonly from Benton, M., 1997, *Vertebrate Palaeontology*, Chapman and Hall. Figure 11.3: simplified from Benton, M., 1997, *Vertebrate Palaeontology*, Chapman and Hall (Figure 9.6d). Figure 11.4: redrawn from Benton, M. and Harper, D., 1997, *Basic Palaeontology*, Addison Wesley Longman. Figures in Table 11.1: redrawn from Black, R., 1979, *The Elements of Palaeontology*, Cambridge University Press (Figure 188). Figure 11.11: redrawn from Prothero, D.R., 1998, *Bringing Fossils to Life*, W.C.B./McGraw-Hill USA (Figure 17.46).

Figure 12.1: redrawn and modified from Benton, M. and Harper, D., 1997, *Basic Palaeontology*, Addison Wesley Longman (Figure 10.8, p. 232). Figure 12.3: redrawn from Andrews, H.N.Jr., 1960, *Palaeobotanist*, 7, 85–9. Figure 12.4: courtesy of Edwards, D., 1970, *Palaeontology*, 13, 150–5. Figure 12.5: redrawn from, Edwards, D.S., 1980, *Reviews of Paleobotany and Palynology*, 29, 177–88. Figure 12.6: redrawn from Andrews, H.N. and Kasper, A.E., 1970, *Maine State Geological Survey Bulletin*, 23, 3–16 (Figure 6). Figure 12.7: redrawn and modified from Eggert, D.A., 1974, *American Journal of Botany*, 61, 405–13. Figure 12.8: based on various sources. Figure 12.9a: redrawn and modified from Bold, H.C., Alexopoulos, C.J. and Delevoryas, T., 1987, *Morphology of Plants and Fungi*, Harper International Edition (Figure 25-16, p. 613). Figure 12.9b: redrawn from Stewart, W.N. and Delevoryas, T., 1956, *Botanical Review*, 22, 45–80 (Figure 9). Figure 12.10: redrawn from Andrews, H.N., 1961, *Studies in Paleobotany*, Wiley & Sons, New York (Figure 11-1). Figure 12.11: redrawn from Delevoryas, T., 1971, *Proceedings of the North American Paleontological Convention*, 1, 1660–74. Figure 12.12: redrawn and modified from Bold, H.C., Alexopoulos, C.J. and Delevoryas, T., 1987, *Morphology of Plants and Fungi*, Harper International Edition (Figure 24-2, p. 584). Figure 12.13: redrawn from Crane, P.R. and Lidgard, S., 1989, *Science*, 246, 675–8. Figure 12.14: courtesy of Sun, G., Ji, Q., Dilcher, D.L., Zheng, S., Nixon, K.C. and Wang, X., 2002, *Science*, 296, 899–904 (Figure 3). All fossil plants from *British Palaeozoic Fossils*, British Museum (Natural History) (Plates 38(1,2,3,5), 39(2,4,5), 40(3)).

Figure 13.1a, b: redrawn from Lipps, J.H., 1993, *Fossil Prokaryotes and Protists*, Blackwell Scientific Publications (Figure 6.2G and 6.2I, respectively, p. 79); Figure 13.1c: redrawn from Wall, D., 1962, *Geological Magazine*, 99, 353–62. Figure 13.2: redrawn and modified from Brasier, M.D., 1980, *Microfossils*, Chapman and Hall, London (Figure 4.2d, p. 23); Wall, D. and Dale, B., 1968, *Micropalaeontology*, 14, 265–304. Figure 13.3a: redrawn from Lipps, J.H., 1993, *Fossil Prokaryotes and Protists*, Blackwell Scientific Publications (Figure 11.3B, p. 171). Figure 13.5: based on various sources. Figure 13.6: modified and redrawn from various sources. Figure 13.7: redrawn from Brasier, M.D., 1980, *Microfossils*, Chapman and Hall, London (Figure 13.1a, p. 90). Figure 13.8: based on various sources. Figure 13.9: redrawn from Brasier, M.D., 1980, *Microfossils*, Chapman and Hall, London (Figure 14.1a). Figure 13.10: based on various sources. Figure 13.11: redrawn from Brasier, M.D., 1980, *Microfossils*, Chapman and Hall, London (Figure 14.10b, p. 134). Figure 13.12: based on various sources. Figure 13.13a–c: based on Brasier, M.D., 1980, *Microfossils*, Chapman and Hall, London (Figure 16.5–7, pp. 157–8); Figure 13.13d: redrawn from McKinney, F.K., 1991, *Exercises in Invertebrate Palaeontology*, Blackwell Scientific Publications (Figure 16.2a, p. 242); Boardman, R.S., Cheetham, A.H. and Rowell, A.J. (eds), 1987, *Fossil Invertebrates*, Blackwell Scientific Publications. Figure 13.14: redrawn and modified from Briggs, D.E.G., Clarkson, E.N.K. and Smith, M.P., 1983, *Lethaia*, 16, 1–14 (Figure 2). Figure 13.15: redrawn from Goudie, A., 1982, *Environmental Change*, Oxford University Press (Figure 2.7A, p. 51).

Hystrichosphaeridium: redrawn from Brasier, M.D., 1980, *Microfossils*, Chapman and Hall, London (Figure 4.3g, p. 26); Tschudy, R.H. and Scott, R.A. (eds), 1969, *Aspects of Palynology*, Wiley-Interscience, New York. *Coscinodiscus*: redrawn from Lipps, J.H., 1993, *Fossil Prokaryotes and Protists*, Blackwell Scientific Publications (Figure 10.5B, p. 159). *Bathropyramis*: redrawn from Brasier, M.D., 1980, *Microfossils*, Chapman and Hall, London (Figure 12.7c, p. 87). *Globigerina and Bolivina*: redrawn from Prothero, D.R., 1998, *Bringing Fossils to Life*, W.C.B./McGraw-Hill USA (Figure 11.7, p. 194). *Beyrichia*: redrawn from Brasier, M.D., 1980, *Microfossils*, Chapman and Hall, London (Figure 14.10c, p. 134). *Cypridina*: redrawn from McKinney, F.K., 1991, *Exercises in Invertebrate Palaeontology*, Blackwell Scientific Publications (Figure 6.7, p. 87). *Cypris*: redrawn from Brasier, M.D., 1980, *Microfossils*, Chapman and Hall, London (Figure 14.7, p. 129). *Bythocertina*: redrawn from Brasier, M.D., 1980, *Microfossils*, Chapman and Hall, London (Figure 14.18c, p. 141).

Figure 14.1: based on Frey, R.W., Pemberton, S.G. and Saunders, T.D.A., 1984, *Bulletin of Canadian Petroleum Geology*, 33, 72–115 (Figure 7). Figure 14.2: based on Frey, R.W., Pemberton, S.G. and Saunders, T.D.A., 1990, *Journal of Paleontology*, 64(1), 155–8 (Figure 1); Brenchley, P.R. and Harper, D.A.T., 1998, *Palaeoecology: Ecosystems, Environments and Evolution*, Chapman and Hall (Figure 5.6, p. 155).

Figure 15.1b: redrawn and simplified from Benton, M. and Harper, D., 1997, *Basic Palaeontology*, Addison Wesley Longman (Figure 4.4). Figure 15.4: redrawn and modified from Benton, M. and Harper, D., 1997, *Basic Palaeontology*, Addison Wesley Longman (Figure 4.7). Figure 15.5: based on various sources.

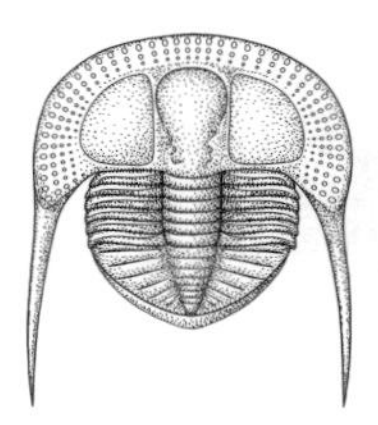

1 Introduction

- Fossils provide information on geological time, evolutionary history, and ancient environments.
- There are two main types of fossil: body fossils and trace fossils.
- Different organisms have different likelihoods of preservation and the fossil record is generally biased towards shelly, marine organisms.
- Fossil Lagarstätten are deposits with exceptionally well-preserved specimens providing an unequaled view of past organisms.

Introduction

Fossils are the remains of organisms and the evidence of past life. They have been part of Earth history for perhaps over 3.5 billion years and have enormous potential for providing information about the geological past. As indicators of past environments and through their ability to define a high-resolution relative timescale, fossils make a vital contribution to the earth and environmental sciences. In the biological sciences fossils are extremely important as they are the only real test for theories of evolutionary process and rate. Furthermore, fossils provide a narrative for life on Earth and reveal how life has adapted to different environments and responded to the challenges of a dynamic system. The natural experiment of life on Earth is currently our only window onto the ways in which life changes in response to external pressure, and also the only window onto how life can change a planet.

In this book information about fossils is arranged in 16 short chapters. This chapter explains what fossils are and gives a broad outline of their uses. Chapter 2 deals with evolution, classification, and the biological basis of change with time. Chapters 3–10 deal with each of the most important invertebrate fossil groups in turn, starting with the simplest, sponges, and finishing with the most closely related to vertebrates, graptolites. Such an emphasis on invertebrates reflects the relative weight given to these groups in most taught courses in palaeontology. Following this are chapters on vertebrates, land plants, microfossils and trace fossils. The two final chapters, 15 and 16, set out the major biological events of the Precambrian and Phanerozoic and sketch the relationship between the evolution of Earth and of its cohort of organisms.

This book is designed to help you access the wealth of palaeontological information. Each chapter is self-contained and pages are arranged in such a way that a particular topic is confined to two facing pages, for example, all of the information on rugose corals is organized on pp. 24 and 25. This means that the book can be used in laboratory classes, where diagrams can be compared with the morphology of real specimens. In addition, putting the information into small compartments means that most sections can be read by themselves, rather than the book needing to be read cover to cover, or a whole chapter at a time. This is why the book is called *Fossils at a Glance*.

What it a fossil?

The word fossil is derived from the Latin *fossilis* meaning an object that has been dug up from the ground. Fossils are the evidence for the existence of once-living animals and plants and may be either the preserved remains of an organism or evidence of its activity.

Types of fossils (Fig. 1.1)

Trace fossils

Trace fossils are the preserved impressions of biological activity. They provide indirect evidence for the existence of past life. They are the only direct indicators of fossil behavior. As trace fossils are usually preserved *in situ* they are very good indicators of the past sedimentary environment. Trace fossils made by trilobites have provided an insight into trilobite life habits, in particular walking, feeding, burrowing, and mating behavior.

Chemical fossils

When some organisms decompose they leave a characteristic chemical signature. Such chemical traces provide indirect evidence for the existence of past life. For example, when plants decompose their chlorophyll breaks down into distinctive, stable organic molecules. Such molecules are known from rocks more than 2 billion years old and indicate the presence of very early plants.

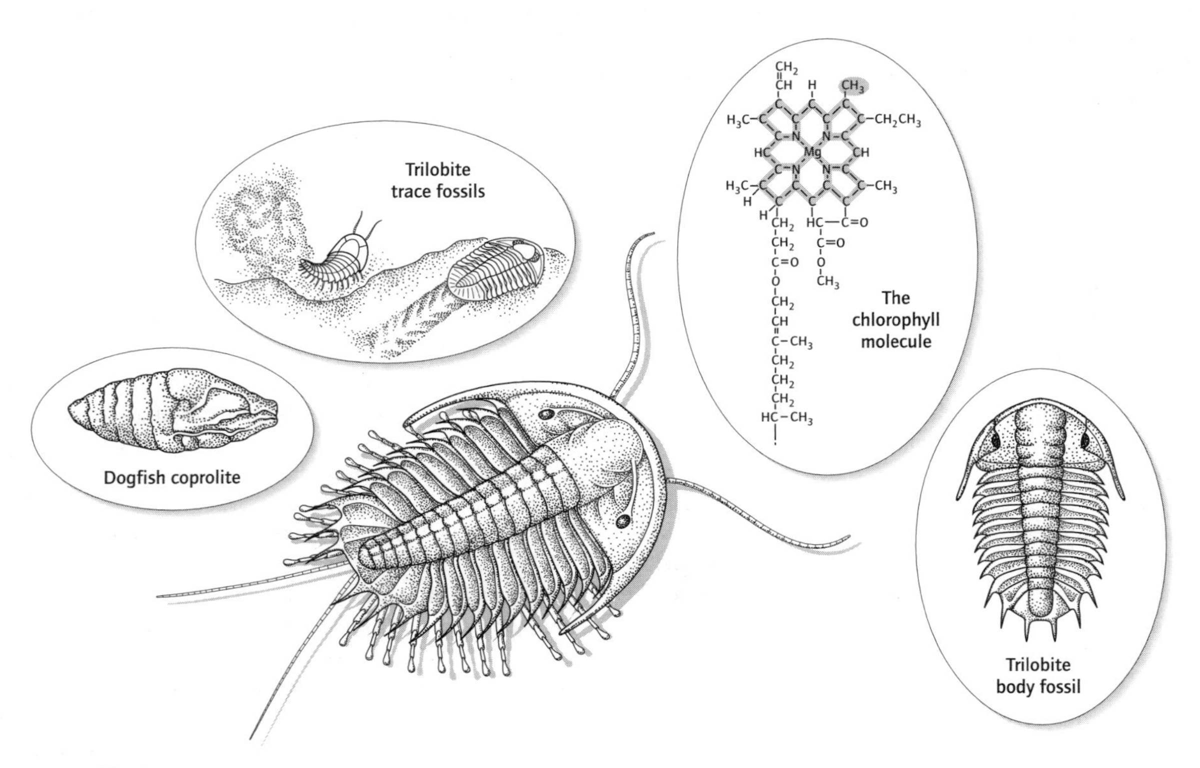

Fig. 1.1 Types of fossils.

Coprolite

Coprolites are fossilized animal feces. They may be considered as a form of trace fossil recording the activity of an organism. In some coprolites recognizable parts of plants and animals are preserved, providing information about feeding habits and the presence of coexisting organisms.

Body fossils

Body fossils are the remains of living organisms and are direct evidence of past life. Usually only hard tissues are preserved, for example shells, bones, or carapaces. In particular environmental conditions the soft tissues may fossilize but this is generally a rare occurrence. Most body fossils are the remains of animals that have died, but death is not a prerequisite, since some body fossils represent parts of an animal that are shed during its lifetime. For example, trilobites shed their exoskeleton as they grow and these molts may be preserved in the fossil record.

Time and fossils

Geological time can be determined absolutely or relatively. The ages of rocks are estimated numerically using the radioactive elements that are present in minute amounts in particular rocks and minerals. Relative ages of different units of rocks are established using the sequence of rocks and zone fossils. Sediments are deposited in layers according to the principle of superposition, which simply states that in an undisturbed sequence the older rocks are overlain by younger rocks.

Zone fossils are fossils with a known relative age. In order for the zone to be applicable globally it must be abundant on a worldwide scale. Most organisms with this distribution are pelagic, that is they live in the open sea. The preservation potential of the organism must also be high, that is they should have some hard tissues, which are readily preserved.

Stratigraphy

The study of sequences of rocks is called stratigraphy. There are three main aspects to this study: chronostratigraphy, lithostratigraphy, and biostratigraphy (Fig. 1.2).

Chronostratigraphy establishes the age of rock sequences and their time relations. Type sections are often established. These are the most complete and representative sequences of rocks corresponding with a particular time interval. For example, the Wenlock Series, which outcrops along Wenlock Edge in Shropshire, UK, is the type section for the Wenlock Series of the Silurian.

A point in a sequence is chosen that represents an instant in geological time and also corresponds with the first appearance of a member of a distinctive series of zone fossils. Relative timescales can then be established with reference to this precise point. These points are called "golden spikes".

The differentiation of rocks into units, usually called formations, with similar physical characteristics is termed lithostratigraphy. Units are described with reference to a type section in a type area that can be mapped, irrespective of thickness, across a wide geographic area.

In biostratigraphy, intervals of geological time represented by layers of rock are characterized by distinct fossil taxa and fossil communities. For example, the dominant fossils in Palaeozoic rocks are brachiopods, trilobites, and graptolites.

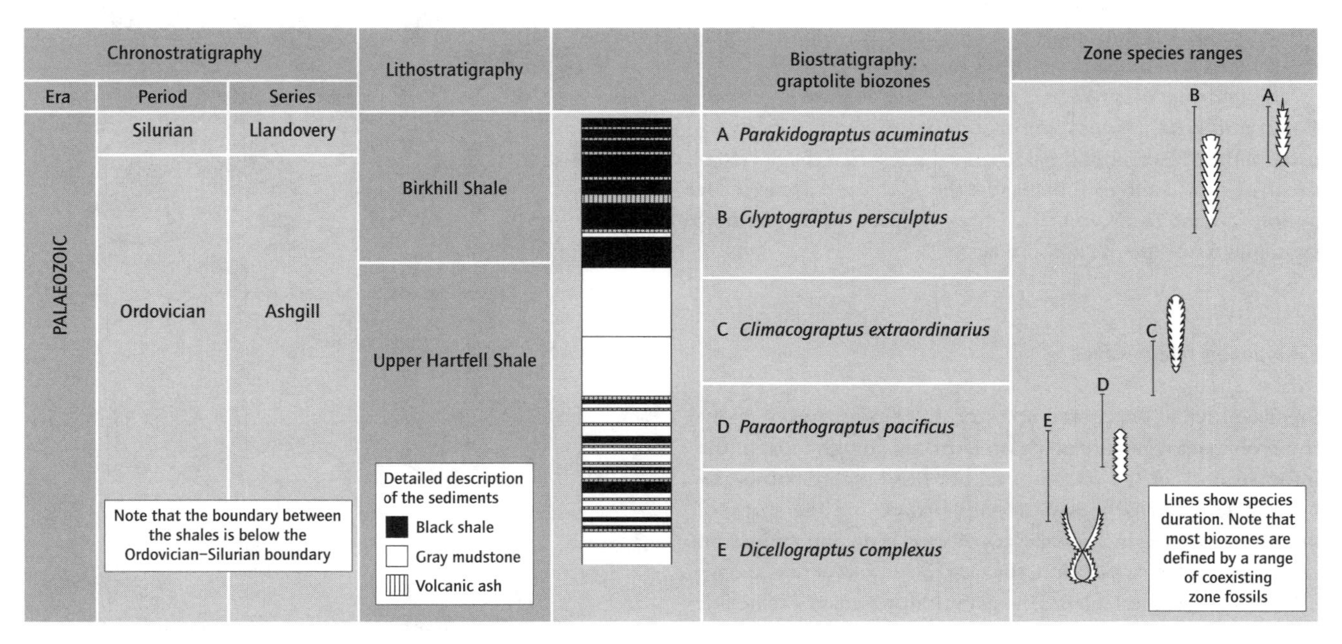

Fig. 1.2 Stratigraphic description of the sequence of rocks that crosses the Ordovician–Silurian boundary at Dobb's Linn, Southern Uplands, Scotland. Geological time is split into different zones depending on the method of analysis. Chronostratigraphy divides the section into two periods. Lithostratigraphic analysis divides the sequence into two shales. Biostratigraphy, as determined by the zone fossils, gives a more detailed description of the sequence.

Information from fossils

Time

As already mentioned, the use of fossil sequences is the most practical method for determining geological time. A species lives for a specific period of time; from its origin to its extinction. Using these vertical time ranges, layers of rock (strata) from different geographic areas can be correlated in time. This technique is called biostratigraphy.

Layers of rocks with distinctive fossil assemblages are called biozones. A good zone fossil should be easy to identify and have a short time range (ideally about 1 million years) (Fig. 1.3).

	Zone fossil
Younger	*Isograptus*
Lower Ordovician	*Didymograptus*
	Tetragraptus
Older	*Clonograptus*

Fig. 1.3 Graptolite zone fossils used to date Lower Ordovician rocks in North America and Scotland.

Palaeoecology

Palaeoecology is the study of ancient organisms and their relationship with the environment (Fig. 1.4). Most palaeoecological studies are based on comparisons with similar living communities. Evolutionary palaeoecology studies the large-scale trends in ecological processes through time.

This information provides insights into ecological processes. For example, large-scale comparisons of the community structures of the Mesozoic and Palaeozoic suggest that the development of new ecological niches in the Mesozoic allowed for a more diverse fauna and that Mesozoic prey developed new strategies to escape the new predators.

Fig. 1.4 Upper Ordovician trilobite–brachiopod community.

Evolutionary relationships

Fossils provide important and accurate information about the evolutionary history of life on Earth. Although most of the information is at the species level the fossil record shows the large-scale patterns of evolutionary change and the shape of life through time, including rates of speciation and extinction (Fig. 1.5). Adaptive radiation, the rapid evolution of new species into new ecological niches, usually follows mass extinction events.

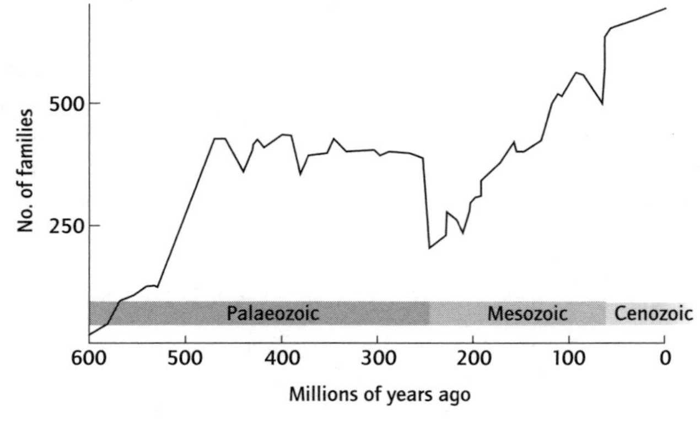

Fig. 1.5 The pattern of mass extinctions shown by the fossil record.

Palaeoenvironment

Fossils provide information about the physical nature of the Earth through time. Photosynthetic production of oxygen increased the amount of the gas in the atmosphere and may have facilitated some of the early radiations of organisms. Climatic conditions may also be determined from palaeontological information. Warm global climates are characterized by extensive developments of calcified cyanobacteria.

Fossils also provide information about the distribution of the continents. Biogeography is the study of the geographic distribution of plants and animals. The fossil record reveals that the patterns of faunal provinces have changed through time, allowing the mapping of the movement of continents (Fig. 1.6).

Fig. 1.6 Two different groups of brachiopods and trilobites are found in the Lower Ordovician rocks of Britain, suggesting that a "barrier", in this case a deep ocean, which has since closed, existed between the two faunas.

Information loss

Habitat

As most material is transported the information regarding the original habitat of the organism is lost. Very rarely, animals and plants may be buried where they lived, in particular burrowing or cemented organisms.

Community structure

The transport of individuals also results in the redistribution of organisms in the fossil record. Displaced fossils can be identified as they are usually physically damaged during transport. Exceptionally, some fossil assemblages are preserved together in place, providing an insight into the structure of ancient communities.

Behavior

The only direct evidence of animal behavior is from trace fossils. As the tracemaker is rarely preserved in conjunction with the trace, most fossil behavior is inferred from comparison with similar living organisms and interpretation of the traces in terms of how they might have been made.

Morphology

Fossilization of the whole organism is extremely rare. Soft tissues usually decay and the skeleton fragments. The living morphology is generally interpreted by comparison with similar living organisms, where possible.

Fossil preservation

The fossil record is incomplete. Most organisms do not fossilize and most fossils are only the partial remains of once-living organisms. Those organisms that do fossilize are usually changed in some way. Most plants and animals are not preserved in their life position and their composition is usually altered.

The study of the history of an organism from its death to its discovery within a rock or sediment is known as taphonomy (Fig. 1.7). After the death of an organism, physical and biological processes interact with the organic remains. This determines the extent to which the organism is fossilized and the nature of the fossil.

The general taphonomic history of a fossil is as follows. After death, the soft tissues of the organism decay. The remaining hard tissues are then transported resulting in disarticulation and possible fragmentation. The broken hard tissues are then buried and are physically or chemically altered. Postburial modifications are termed diagenesis. This sequence of events results in a major loss of information about the organism and its life habit.

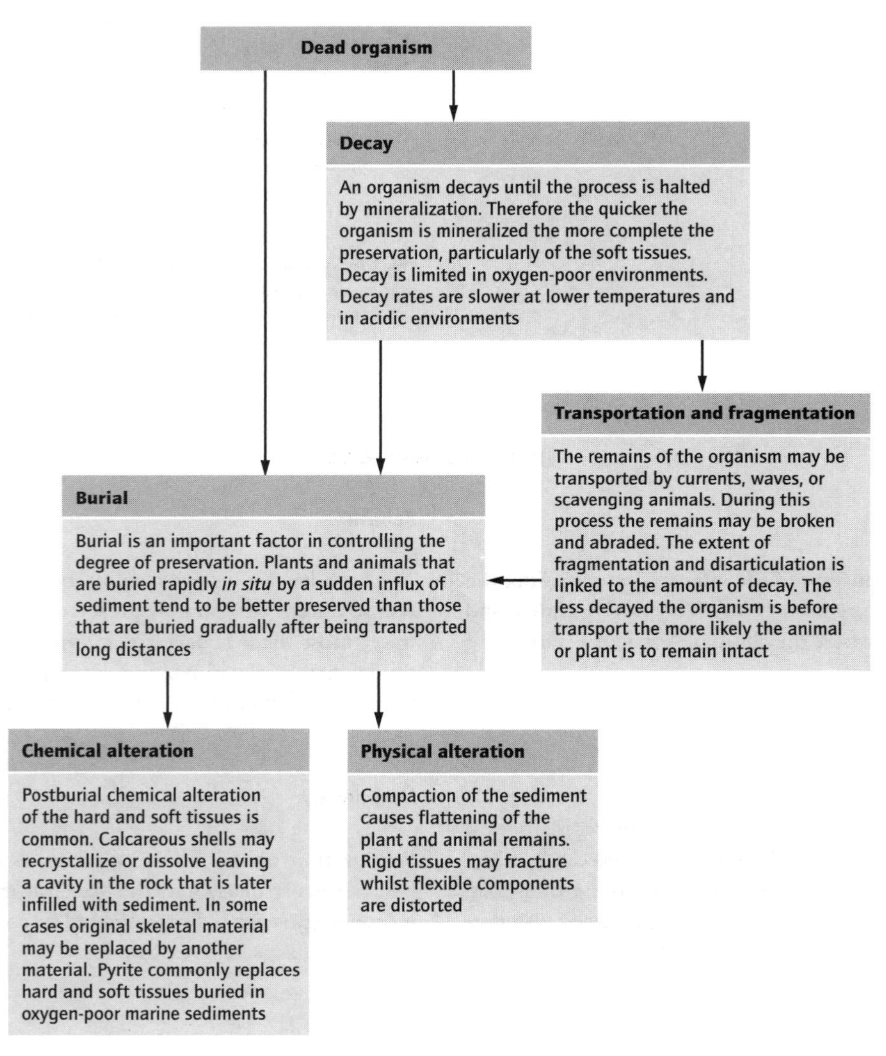

Fig. 1.7 The process of fossilization (taphonomy).

Biases in the fossil record

The fossil record is extremely selective. The term preservation potential is used to describe the likelihood of a living organism being fossilized. Organisms with a high preservation potential are common fossils. The nature of their morphology and the environment in which the fossil lived are important factors in determining whether it will be preserved. These inherent biases skew our view of past life. In general, the fossil record is biased towards the following:

- organisms with tissues resistant to decay;
- marine organisms;
- organisms living in low energy environments.

Organisms with tissues resistant to decay

Organisms with body parts that do not decay easily are more likely to be preserved in the fossil record than soft-bodied animals. In vertebrate animals the teeth and bones are the most commonly fossilized components. Invertebrates often have shells and carapaces that are not prone to decay. The woody tissues of plants, plant spores and pollen are the most likely plant parts to be preserved in the fossil record.

The shells of most common invertebrates are formed from calcium carbonate in the form of calcite or aragonite. Aragonite may be converted to calcite during fossil diagenesis. This can be identified by a change in the shell crystal structure from layers of needle-like crystals to large blocky crystals. Some invertebrates have skeletons composed of silica, for example sponges, that are preserved in the fossil record. The skeleton (or rhabdosome) of graptolites was composed of collagen, a protein which is extremely durable and resistant to decay. Animals with exoskeletons molt as they grow, increasing the number of potential fossils. Plant material is particularly prone to decay, although the woody tissues that form the stem and leaves, together with spores and pollen that have a resistant waxy coating, may be preserved in the fossil record.

The skeletal structure of an organism determines the completeness of the preservation. Hard tissues that are in the form of a single component are more likely to be preserved whole in the fossil record, for example the shells of gastropods or ammonites.

Marine organisms

Marine organisms are more likely to be preserved than those living on land. On land there is more erosion and less deposition of sediment and consequently less opportunity for burial. Terrestrial plants and animals living close to depositional areas, for example by the side of lake, have a greater preservation potential than those living in areas of net erosion such as uplands.

The nature of the substrate that the organism inhabits does not seem to have an effect on the preservation potential of a marine animal. However, its ecology does affect the likelihood of a marine animal being fossilized. Sedentary animals, filter-feeders, and herbivores are more commonly preserved in the fossil record than carnivorous animals. Sedentary animals, like corals, tend to be heavy and robust whilst active predators have more lightly constructed skeletons. In addition, mobile animals can escape from burial by sediment.

Organisms living in low energy environments

In low energy environments the mechanical processes, such as currents, waves, and wind, that destroy plant and animal remains are less intense. Therefore organisms living in these environments are more likely to be preserved. However, this can be an oversimplistic view since organisms living in high energy environments may have more developed and more durable skeletons, thus increasing their preservation potential, or they may be buried more quickly and hence avoid postmortem damage.

Exceptionally preserved fossils

Remarkable fossil deposits are known as fossil Lagerstätten. Lagerstätten is a German word that is applied to deposits of economic importance. The term fossil Lagerstätten is used to describe fossiliferous formations particularly rich in palaeontological information. There are two types of fossil Lagerstätten: Konzentrat-Lagerstätten and Konservat-Lagerstätten (Figs 1.8 and 1.9). Occurences where the number of fossils preserved is extraordinarily high are termed Konzentrat-Lagerstätten or concentration deposits. In Konservat-Lagerstätten, or conservation deposits, the quality of preservation is exceptional, soft tissues are fossilized, and the skeletons are articulated. Konservat-Lagerstätten are a rich source of palaeontological information. Preservation of the soft tissues helps explain the palaeobiology of extinct organisms and the preservation of an entire community provides an insight into the structure of ancient ecosystems. Konservat-Lagerstätten can be considered as "preservation windows" that provide an exceptional view of past life.

Konservat-Lagerstätten generally form in environments that are hostile to life or in environments with very high sedimentation rates. Carcasses may be transported into hypersaline or anoxic lakes that are devoid of scavengers. Such occurrences produce stagnation deposits. Rapid burial also minimizes the effect of scavengers. This can occur in deep marine environments where turbidity currents may suddenly deposit large quantities of sediment or in a delta where large volumes of material are being discharged into the sea. These deposits are called obrution deposits. Konservat-Lagerstätten are also associated with conditions that cause instant preservation. These situations are known as conservation traps and include insects preserved in amber (fossilized tree resin) and animals trapped in peat bogs.

Konzentrat-Lagerstätten
Concentration deposits
Formations with high abundance of fossils

Konservat-Lagerstätten
Conservation deposits
Formations with exceptionally preserved fossils

Stagnation deposits
Formed in anoxic or hypersaline conditions

Obrution deposits
Formed by rapid burial

Conservation traps
Instant preservation in restricted situations

Fig. 1.8 Classification of fossil Lagerstätten.

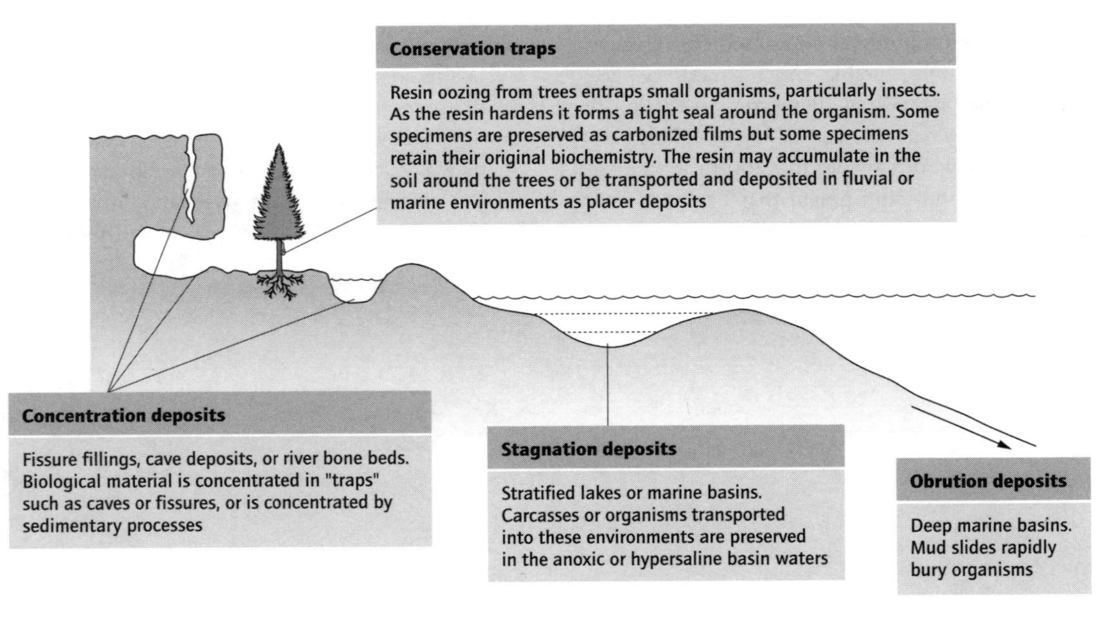

Fig. 1.9 Types of fossil Lagerstätten.

Examples of Konservat-Lagerstätten

Baltic amber, Tertiary, Russia

Amber is fossilized plant resin. As the name implies, Baltic amber is abundant along the shores of the Baltic Sea, particularly around the Samland Promontory of Russia.

In the early Tertiary, forests of the extinct tree *Pinus succinifera* flourished on a land mass south of the Samland region. During the middle Tertiary the area was flooded and the resin from the trees was washed out and redeposited in marine sediments in the Samland area. These sediments have been reworked and the amber was subsequently redeposited in areas along the shores of the Baltic Sea. As amber has a low density it can be carried by water in suspension and is generally deposited in low energy environments such as lakes, submarine basins, and estuaries.

Around 98% of the Baltic amber biota are flying insects. Diptera, two-winged flying insects, dominate the fauna of the Baltic amber, accounting for approximately half of the organisms (Fig. 1.10). Extremely rare mammal hairs, an almost complete lizard, snails, and bird feathers account for the remaining 2%.

The Baltic amber fossils are important as they show the morphology of flying insects in extremely fine detail and also provide information on the dispersal and development of these insects and the climatic conditions in which they lived.

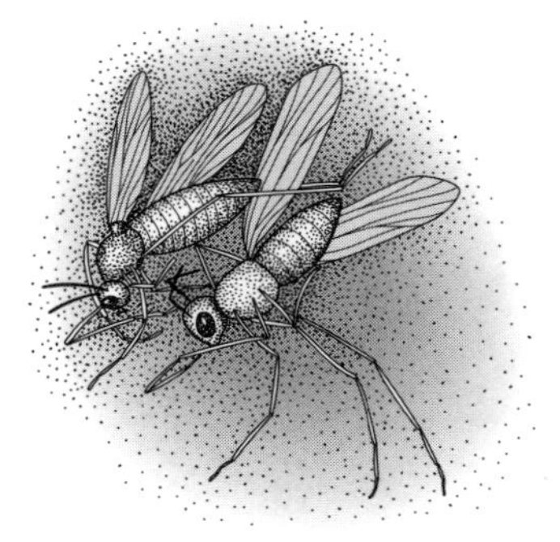

Fig. 1.10 Flies preserved in amber from the Baltic.

Solnhofen Lithographic Limestone, Jurassic, Bavaria

Most famous for the preservation of the feathered dinosaur *Archaeopteryx*, the Solnhofen Lithographic Limestone (Jurassic) outcrops over a wide area of Bavaria. The limestones are buff colored, fine grained, and extraordinarily regular, forming laminar beds traceable over tens of kilometers. They were deposited in a series of lagoons formed behind reefs. High evaporation rates and limited water exchange with the open sea caused the lagoon waters to stratify with the more dense saline waters forming a hostile bottom environment.

Over 600 species of animals and plants are preserved in the limestones (Fig. 1.11). Most animals were pelagic. Only a few benthic organisms are known from the area. It is believed that most of these animals were swept in from the open sea. Some were able to live in the less saline upper waters for short periods. Terrestrial animals and plants may have been washed in during rainy seasons and insects blown into the lagoons by the wind.

Fig. 1.11 The most common macrofossil in the Solnhofen Limestone: the crinoid, *Saccocoma* (diameter 5 cm).

Burgess Shale, Middle Cambrian, British Colombia

The Burgess Shale is one of the most celebrated Fossil-Lagarstätten, providing a unique insight into the nature of Cambrian life and the evolution of multicellular animals. Over 65,000 specimens have been recovered from a few small quarries in the Canadian Rockies.

Arthropods and soft-bodied worms are the most common fossils but there are a number of animals which show unusual body plans that have no modern counterpart. These bizarre animals show the extremely high degree of adaptive radiation that took place in the Cambrian explosion of life.

The exceptional preservation of these organisms is a result of a submarine landslide that transported them from the shallow bank where they were living into deep anoxic waters. The organisms were rapidly buried in fine mud that prevented their decay. Soft tissues were then replaced by silicate minerals preserving them in very fine detail.

Glossary

Adaptive radiation – evolutionary response to large-scale environmental change. This results in the formation of new ecological niches that can be occupied through the adaptation of previous generalists.
Amber – fossilized tree resin.
Benthic – living on or in the sea floor.
Biostratigraphy – stratigraphy based on fossil content.
Biozone – layer of rock characterized by fossil content.
Chlorophyll – green pigment found in plants.
Chronostratigraphy – stratigraphy based on geological time.
Cyanobacteria – microorganisms with chlorophyll that produce oxygen on photosynthesis.
Diagenesis – physical and chemical processes that operate on sediments after burial.
Exoskeleton – external skeleton of an animal.
Golden spike – physical point in a section equivalent to an instant in geological time marking the base of a stratigraphic unit.
Infaunal – living within the sediment.
Konservat-Lagarstätten – deposits containing exceptionally preserved fossils.
Konzentrat-Lagarstätten – deposits containing numerous fossils.
Lagarstätten – deposits containing numerous and/or exceptionally preserved fossils.
Lithostratigraphy – stratigraphy based on rock characteristics.
Pelagic – living in open water (floating or swimming).
Photosynthesis – biological process in plants that captures light energy and converts it into chemical energy.
Plate tectonics – theory that the Earth's crust is formed of moving plates.
Taphonomy – study of the process of fossilization.
Taxon (plural taxa) – general term for any formal grouping of plants and animals.
Zone fossil – fossil species that is indicative of a particular unit of geological time.

2 Fossil classification and evolution

- Classification, or taxonomy, enables organisms to be categorized, either in order to identify them or in order to understand their evolutionary history.
- The diversity of organisms on Earth is a reflection of the process of evolution by natural selection.
- Microevolution describes the origin of new species. There are two main models of speciation, by gradual increments or sudden steps.
- Macroevolution deals with the origin of larger groups than a species, and describes trends in the fossil record.

Introduction

Life probably began with a single living organism, while diversity today is estimated at between 2 and 20 million species. This increase in diversity needs to be quantified and understood. The first of these aims is achieved by classifying organisms, the second by the study of evolution.

The most common unit of classification, or taxonomy, is the species. This is defined as a potentially interbreeding group, whose individuals cannot breed with other organisms outside the group. A species may be considered as the sum of all the genes that could be shared by its members. In practice, species are usually defined by their physical similarity to one another. Modern techniques of DNA and enzyme extraction have allowed some classification using biochemical characters.

Species are grouped together into genera, genera into orders, and so on, until all life is linked within a series of nested sets. The aim of taxonomy is usually to produce a "natural" classification, where these nested sets reflect the history of each species through time, and display the evolutionary relationships of that species. Sometimes a different aim in classification is to make identification simple, or to produce a working description of diversity in advance of an understanding of the evolution of a group. In this case the classification is useful but artificial (Fig. 2.1). Traditional methods of classification are supplemented by cladistic techniques that produce diagrams showing the simplest possible relationships between organisms given the characters they share.

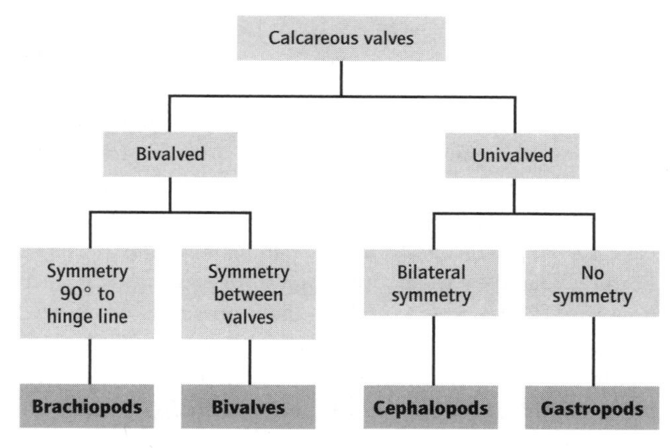

Fig. 2.1 An artificial classification separating major classes of molluscs and brachiopods using the minimum number of characters.

Natural classifications reflect the ways in which species have changed with time, and the ways in which larger groups, such as orders, have arisen. Understanding how these events occur is the realm of evolutionary studies, and depends on two vital elements. One is an appreciation of change being driven by natural selection, as suggested by Darwin. The other is an understanding of the genetic basis for such change. In addition to investigating the mechanisms of change, it is important to study rates of change. As species generally live for between 4 and 10 million years, this is something best done by using the fossil record.

Taxonomy

The two purposes of modern taxonomy are to provide a working classification for life, that allows species to be identified, and to map out the evolutionary relationships of a group of organisms. This kind of mapping process is the study of phylogeny.

All organisms are classified as belonging to a series of nested groups, the smallest unit of which is the species. Species are linked together into genera (singular, genus), genera into orders, orders into classes, classes into phyla (singular, phylum), and phyla into kingdoms. Sometimes intermediate stages are identified, such as subclasses or superfamilies. An organism is usually given two names, the first of which begins with a capital letter, and is the name of its genus. The second begins with a lower case letter and is the name of the organism's species. Both of these names should be written in italics, or underlined. Higher up the taxonomic ladder plain text is used. An example of this hierarchy, for the ammonite *Asteroceras obtusum*, is shown in Fig. 2.2.

All member of a species share many common features. In turn, all species within a genera will have a significant number of shared characters. As the hierarchy develops, the number of shared characters gets fewer, but their importance, especially in terms of early growth, symmetry, and body form tends to increase. In an ideal example, the taxonomic hierarchy of an organism will define its phylogeny; that is, will be a map of its evolutionary history. All species within a genus should share a common ancestor. All genera in an order should share a more distant common ancestor. The phylogeny of some groups, for example bivalves, is well understood. However, in other groups, such as graptolites, establishing the phylogeny of most members of the group is problematic, and their classification is partly an artificial one.

Category	Taxon
Kingdom	Animalia
Phylum	Mollusca
Class	Cephalopoda
Subclass	Ammonoidea
Order	Ammonitida
Genus	*Asteroceras*
Species	*Asteroceras obtusum*

Fig. 2.2 Taxonomic description of the ammonite *Asteroceras obtusum*.

Cladistics

The points at which groups of organisms have separated are difficult to determine. Cladistic analysis attempts to identify these branching points and establish evolutionary relationships between a diverse group of organisms. Closely related groups have shared characters termed "shared derived characters". Features shared by the entire group of organisms are termed primitive. Characters may be called primitive or derived depending on the level of classification. For example, feathers would be considered a derived character in the vertebrates but a primitive character in the birds. Shared characters are shown on a branching diagram called a cladogram (Fig. 2.3). Each shared character is evaluated and a decision is made, by reference to a more distant relative (termed an outgroup), as to whether the state is primitive or derived. Cladograms are then constructed using a computer program. The program will produce a number of possible cladograms. These are then evaluated by the cladist.

Although superficially similar to evolutionary trees, cladograms show only the relationships between organisms, not their distribution in time. Taxa are shown on the tips of each branch. Groups with a unique common ancestor are known as sister groups. Each branch is supported by a shared character from a hypothetical common ancestor. Ancestors are rarely physically identified. This is partly because of the incompleteness of the fossil record but also because by focusing on shared evolutionary characters, rather than searching for ancestors, the pattern of relationships between groups of organisms has been discovered.

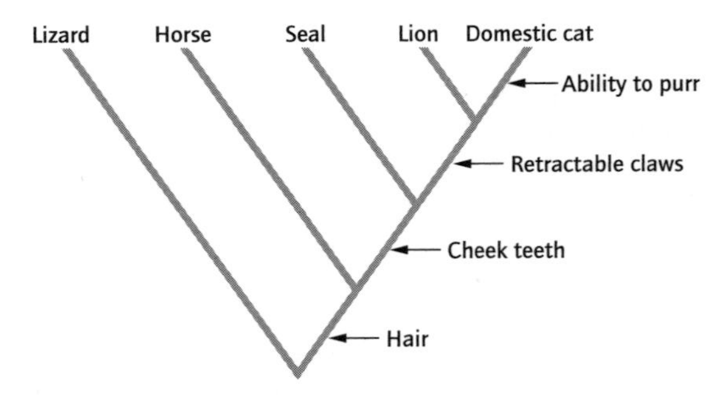

Fig. 2.3 A simplified cladogram. The characters next to the arrows are the shared characters that define the branching points. This cladogram shows that lions and domestic cats share a common ancestor and are therefore sister groups.

Evolution by natural selection

The idea that evolution takes place by natural selection was proposed by Darwin in 1859. In essence his argument can be put as follows:

1 Species have the potential to produce more offspring than the environment can sustain.

2 Individuals within a species have to compete for scarce resources.

3 Within a species there is a degree of variation between individuals and some of this variation is inherited.

4 Those individuals possessing characteristics that give them a competitive advantage will be more likely to survive and to reproduce. Their offspring have a better chance of survival.

Over time, the process of natural selection of those organisms best "fitted" to their environment leads to changes in the overall set of characters possessed by the species. This provides a simple but powerful tool for changing morphology with time, that is, for evolution.

Several fascinating questions are generated by a consideration of Darwin's idea. First, how is variation generated and inherited? Second, what drives selection – the physical environment or the interaction of an organism with other species? Third, at what rate does evolution occur – is it continuous or episodic? The first of these was the one that most disturbed Darwin. Only after his death was the genetic basis of life discovered and the answer to this question found.

The genetic basis for evolution

The information to build organisms, which is transmitted from parent to offspring, is held on DNA (deoxyribonucleic acid) or RNA (ribonucleic acid), within cells. In advanced organisms DNA is folded into the cell nucleus. Strands of DNA are made up of sequences of four molecules: adenine, cytosine, guanine, and thymine. These molecules are organized into groups of three forming a triplet code. Discrete sections of this code that perform particular functions are known as genes. The entire set of instructions is called the genome. Within the DNA there are long sequences of unused code, punctuated by shorter sequences that contain the instructions for producing the organism. In addition, controlling genes regulate the timing and level of expression of particular genes.

Mutation is the change in the sequence of molecules that make up the DNA. Mutations can be caused by imperfections in cell division, environmental changes, such as UV radiation, and toxic chemicals. Changes in the make up of the chromosomes can also be produced naturally through sexual processes. Changes in the control genes or production of too many copies of all or part of the genome can radically affect the form of the organism.

An unexpected bonus in the study of genetics has been the realization that each evolutionary step in the history of an organism is recorded somewhere within the DNA of each living member of the species. However, geneticists have yet to decode the bulk of this story.

In addition, changes in genes with time provides one way to measure the passage of time during the evolution of organisms. This measure of time is known as the molecular clock. Some parts of the genome appear to have changed rapidly with time, whilst other sequences have changed more slowly (these are called conservative gene sequences). An appropriate choice of gene is crucial in measuring time accurately. Conservative sequences can be used to estimate the timing of divergence of major groups of organisms a long time ago, while more rapidly mutating systems can be used to estimate the timing of divergence of species a short time ago. Using multiple gene sequences from the groups being compared increases the chance of being right, but also frequently introduces wide ranges into estimates of evolutionary timing. For example, the origin of multicelled animals is dated from molecular evidence as having occurred between 1,600 and 800 million years ago. In addition, molecular clocks need to be calibrated against known evolutionary events observed in the fossil record, and these may be inadequate tie-points if the fossil record of a group is poor.

Microevolution

The evolution of new species is known as microevolution. A species is defined by all of its potentially interbreeding organisms, and by their shared gene pool. However, in practice species are made up of subunits, such as herds or flocks. While these populations may interbreed with adjacent groups, they are highly unlikely to breed with those animals living a long way away, and in practice most breeding will go on within the group. This offers a method by which new species can arise, as groups are separated from each other, either by distance or by changes in behavior. In this view, microevolution is largely the study of how groups of organisms interact with and migrate through their environment. There appear to be two ways in which new species can emerge, by allopatric or sympatric speciation.

In allopatric speciation a group of organisms becomes physically separated from the rest of the population, perhaps by the appearance of a physical barrier between them, such as a new seaway. The isolated population will experience different environmental and biological stresses to those experienced by the parent, so that a different set of characteristics will be favored. Over time, these new characteristics will become so clearly differentiated from the parental set that the group ceases to be able to interbreed, even if the barrier to migration is removed. A new species has been created. As the collection of characters carried by the isolated population is much smaller than that carried by the parent population, recessive genes are more likely to have an expression. The move towards a new species may be "kick started" simply because the isolated "sample" is not representative of the species as a whole. Alternatively, even if the isolated population is large enough to have a very similar gene pool to the species as a whole, the new conditions it experiences in its new location may be the cause of speciation.

In sympatric speciation, a species splits into two without part of the population becoming physically isolated. Instead, isolation may be achieved by changes in behavior, food, or habitat preference. Again, because the selection pressures acting on the two populations will be different, they will become physically different over time, and will eventually be incapable of interbreeding.

Species that remain as single geographic and behavioral units may still evolve as the conditions they encounter change. In this case, while no new species are created, shape changes will be observed in the fossil record of such a group. When studying such fossils, taxonomists eventually define a point where such change makes it appropriate for the species to be given a new name, on the basis of its new characteristics. The parent species appears to have become extinct, and the daughter species to have originated, but this is actually an artifact of the need to classify organisms into discrete groups. It is sometimes known as pseudoextinction.

The rate at which new species appear has been a major research topic in palaeontology for many years. It was originally assumed, from a first reading of Darwin's ideas, that evolutionary change should be continuous, but slow and steady. Species were expected to change morphology continuously through time, so that eventually a new species, or several, would emerge from the parent population. This idea is known as phyletic gradualism.

However, this steady rate of change is rarely preserved in the fossil record. This may be an artifact of the incompleteness of this record. In most environments the preserved sequence of fossils represents a series of "snap shots" of the population at widely separated intervals of time. These intermittent records tend to produce a jerky" effect on any measurements of shape change. However, in those rare sequences of rocks where deposition is rapid and continuous, such as in some deep lakes, this "jerkiness" still seems to be recorded.

This observation has led to the development of an alternative hypothesis relating to the rate of evolution, known as punctuated equilibrium. This model suggests that species tend to be stable for considerable periods of time, and that daughter species evolve suddenly, in sharp bursts of morphological change. This model effectively implies that the selection pressure experienced by a species is not constant, but episodic. The physical separation of one population from the rest of the species would suddenly change selection pressures, as would a sudden change of climate.

These two models of evolutionary rate are clearly end members of a possible continuum, and both may occur in some situations. Arguments about rate yield insights into the process of evolution, without altering its plausibility.

Macroevolution

The study of the appearance of new groups of species, of new genera, orders, families, and so on, is the province of macroevolution. At this level, the unit of selection is the species, rather than the individual, and the rate of production of new species replaces the rate of production of offspring. An example of macroevolution is the appearance of dinosaurs from thecodonts in the Triassic, or the evolution of birds from saurischian dinosaurs during the Jurassic period (Fig. 2.4).

New genera, orders, families, or even phyla do not appear evenly spread through the fossil record. Instead, there appear to have been some periods of great taxonomic innovation, such as the period around the Precambrian–Cambrian boundary, where many new groups evolved. When such new groups appear, they radiate quickly into a wide variety of forms. This early, explosive phase is known as adaptive radiation. The later history of a taxonomic groups seems to be one of reduced innovation.

The problem of how the large-scale change in morphology needed to generate such new groups operates has been partly solved by an increased understanding of how genes work. We now know that genes have a hierarchy of function, with some genes encoding the production of particular molecules and others controlling the timing and duration of activity of these genes. Changes to regulatory genes can lead to significant changes in morphology. A change in the timing of sexual maturity, for example, can change the whole shape of a reproducing organism, or a change in the instructions controlling the arrangement of segments in arthropods can radically alter the resulting animal.

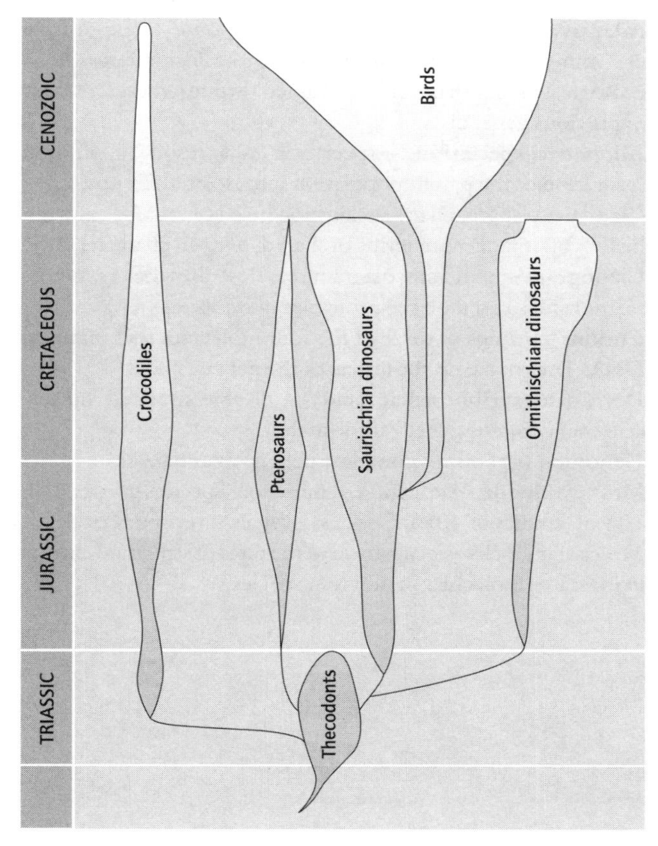

Fig. 2.4 Evolutionary relationships between birds, dinosaurs, pterosaurs, and crocodiles as shown by the fossil record.

Glossary

Adaptive radiation – evolutionary response to large-scale environmental change. This results in the formation of new ecological niches that can be occupied through the adaptation of previous generalists.
Allopatric speciation – speciation as a result of physical separation of the parent population into isolated groups.
Cladistics – approach to classification based on the evolutionary history of organisms in terms of shared, derived characteristics.
Cladogram – branching diagram that describes the taxonomic relationships based on the principles of cladistics.
Codons – groups of three of the four molecules that make up DNA. They make up the letters of the genetic code.
DNA (deoxyribonucleic acid) – double-stranded nucleic acid that contains genetic information.
Genome – organism's complete genetic material.
Mass extinction – extinction of numerous species (greater than 10% of families or 40% of species) over a short time period.
Molecular clocks – comparison of the rates of structural change in indicator molecules in different species.
Mutation – change in the DNA of genes that ultimately causes genetic diversity.
Natural selection – differential selection of organisms. Favorable adaptations are incorporated in the genome of the next generation, increasing its competitive ability.
Phyletic gradualism – theory that evolution is gradual at a more or less constant rate. Speciation occurs as part of this gradual change.
Phylogeny – complete evolutionary history of a species or a group of related species.
Punctuated equilibrium – theory that periods of evolutionary equilibrium are interrupted by episodes of rapid evolutionary change.
Species – population of organisms that has similar anatomical characteristics and interbreeds.
Sympatric speciation – speciation as a result of changes in lifestyle preferences within a population.

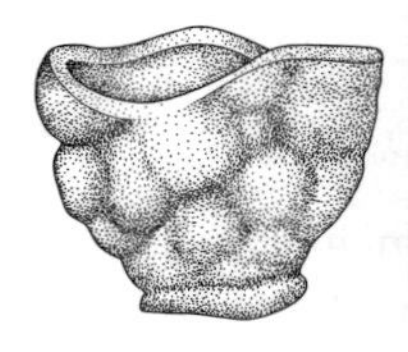

3 Sponges

- Sponges are the simplest type of multicellular organisms.
- Some living sponges can survive being pushed through a sieve, which separates the individual cells, and will reform into a new sponge.
- Sponges have been important in reef environments throughout the Phanerozoic, especially during the Devonian and Permian.
- Sponges have been relatively unaffected by the main mass extinction events.

Introduction

Sponges and their relatives are the most primitive multicellular animals. They are composed of collections of cells that cannot live separately for long, and they lack tissues or integrated sensory functions of any kind. Their internal structure is very simple, and tends to be fractal and to lack symmetry. The position of a cell on a sponge is not fixed, and if the structure is damaged, even pushed through a fine sieve, the sponge will reform. Broken fragments of any size can regenerate into complete organisms.

Sponges build a broadly cup-shaped body, secreting organic material, calcite, aragonite (both $CaCO_3$), silica (SiO_2), or a mixture of these materials. Skeletal material is commonly secreted as individual spicules, sometimes fused together. Water is filtered through the walls of this porous structure, using cells with a whip-like flagellum to generate currents. Waste water is then exhaled through the main aperture of the cup. All sponges are filter-feeders and have a benthic habit. Most of the 10,000 or so living species are marine, but some live in fresh water and a few can survive intermittent exposure to air.

Sponges form the phylum Porifera. Within this phylum, three classes of sponge are identified, based on their skeletal composition (Table 3.1). These classes probably represent convenient classification "baskets" rather than real evolutionary units. It seems that different groups of sponges acquired skeletal elements of different compositions more than once, so composition is not a reliable indicator of their evolutionary relationships.

Two main fossil groups are known, in addition to the groups outlined above. These are the stromatoporoids and the archaeocyathids. Stromatoporoids were major Palaeozoic reef-builders, and are now thought to belong with the common sponges, the demosponges. Archaeocyathids seem to have been an independent "experiment" in forming simple organisms, and may be unrelated to sponges by descent. They evolved in the early Cambrian and were completely extinct by the end of that period. However, their similarity of form and complexity suggest that they can be considered as having been like sponges in most main respects.

All of the main groups of sponges evolved in the Cambrian and they have been part of the marine benthos since that time. In general, calcareous sponges favor shallow water environments, especially high energy areas. Hexactinellid sponges are more common in deeper water and are now found down to abyssal depths. Fossil sponges are relatively rare, at least as complete fossils. Their spicules tend to disaggregate after death and to be preserved separately. They are most easily seen in thin section, where their frequency suggests that sponges were historically more important than at the present time. The exceptions to this poor preservation are the relatively few forms of reef-building sponges, which have left behind dense biological build-ups of calcite.

Table 3.1 Living groups of sponges and their skeletal composition.

Sponge class	Skeletal composition
Calcarea	Calcite spicules or cup walls
Demospongia	Spongin or silica skeletons, sometimes with a calcareous base
Hexactinellida	Silica spicules with sixfold symmetry

Sponge morphology

Sponges are characterized by four important cell types. Archaeocytes are cells shaped like amoebae, able to move within the colony and lacking a fixed shape. These cells are feeding cells and can also change into another cell type if required. Sclerocytes secrete mineralized elements of the skeleton, while spongocytes secrete the organic parts of the skeleton. Choanocytes are the cells that generate feeding currents through the sponge. They have a funnel-shaped end, with a long flagellum, or whip-like filament, extending through the funnel and into the water beyond. The movement of many flagellae, from many choanocytes, helps to move water through the colony.

Sponges have a simple body-shape, characterized by their functional needs as filter-feeding organisms (Fig. 3.1). They construct a skeleton with narrow openings, called ostia, through which water can enter and a broad opening through which water leaves, called the osculum. This has the hydrodynamic effect of facilitating a good flow rate with less work from the flagella-bearing cells. Feeding occurs in the walls of the colony. In the simplest case, archaeocytes line apertures in the wall of the cup-shaped sponge (this is known as ascon-grade organization). In more complicated morphologies, the feeding and water-moving cells are arranged in multiple chambers, linked to a common central area or paragaster (this is the sycon grade of organization). Most commonly, networks of such chambers are developed within the thick wall of the sponge and are linked to the paragaster by a series of canals (the leucon grade of organization).

Calcareous sponges grow skeletons composed of spicules scattered through an organic matrix or of solid calcite. Spicule type is very varied within the group. Demosponges, the most common modern group, usually have unmineralized skeletons, but can produce siliceous spicules, or calcareous bases to their colony. Hexactinellid sponges produce skeletons from a series of six-rayed spicules with each spine arrayed at 90° to its neighbors. These are formed into skeletal frameworks with cubic symmetry. Spicules are the most commonly preserved evidence for the presence of sponges in fossil communities. They can be identified easily in thin section.

Sponges are attached to the sea bed by a holdfast that may resemble roots or a series of fine hairs. These can attach the sponge to hard substrates, such as coral reefs and boulders, or to sand grains and abyssal muds.

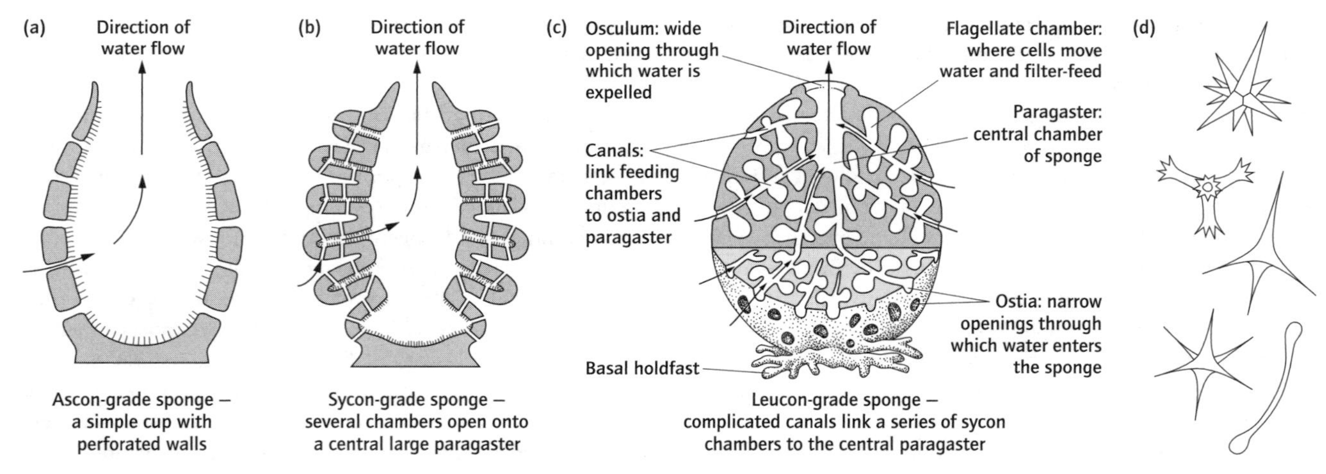

Fig. 3.1 The main features of sponge morphology. Sponges of each of the three main classes grow in broadly similar shapes that can be divided into three grades of internal organization. (a) Ascon-grade organization. These are usually small sponges, less than 10 cm in diameter. (b) Sycon-grade organization. (c) Leucon-grade organization. These complicated sponges can grow much larger than simpler forms, sometimes reaching 50 cm or more in diameter. (d) Sponge spicules, about 1 mm long. Hexactinellid spicules always have six points.

Sponges as reef-builders

The best preserved fossil sponges tend to be reef-formers, and the group has played an important role in building or colonizing reefs through the Phanerozoic. All reef-building sponges had a predominantly calcareous skeleton. Archaeocyathids (Fig. 3.2) evolved into some of the first reef-formers, during a brief period in the early Cambrian. They were small forms, generally around 10 cm in height, with a cup-like shape. Modular variations on this basic plan allowed them to increase in size and form frameworks for reefs. They were mainly tropical and inhabited water less than 30 m deep. Stromatoporoid reefs were common during the Silurian and Devonian, and brief but important intervals of sponge-reef formation occurred in the Permian, Triassic, and Jurassic. In general their importance has declined during the Mesozoic and Cenozoic, possibly linked to the rise of colonial corals with symbiotic algae. However, they were still important members of reef communities throughout this time, especially inhabiting caves and overhanging areas of reefs. These cryptic environments are well preserved in fossil reefs and yield a wealth of calcareous sponge species.

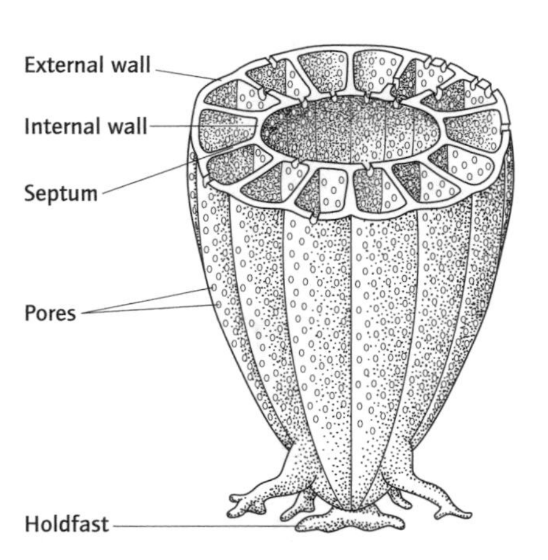

Fig. 3.2 A generalized archaeocyathid. Note the double walls, with septa in between, and the abundant perforations of all skeletal elements. An individual archaeocyathid was typically 5–20 cm in height.

Most modern reefs thrive in nutrient-poor regions, such as around mid-ocean islands. This is because the symbiotic algae of modern corals form a plant base to the food chain. However, the bulk of geologically recognized reefs probably grew in areas where nutrients were less limited because the framework organisms were filter-feeders, and most sponge reefs fall into this category. The exception to this general rule are some Lower Palaeozoic stromatoporoid reefs, leading to the suggestion that this group may also have had symbiotic algae within their cells.

Biogenic silica

Siliceous sponges were the main biological secretors of silica during the Cambrian, when they were mainly confined to shallow water. They formed the dominant biotic flux for this important geochemical cycle. However, since the Cambrian, two important factors have moved the site of this important flux from shallow water to deep water. These are the change in habitat of siliceous sponges and the evolution of plankton-building siliceous skeletons, the radiolarians and diatoms.

The geological record of siliceous sponges is poor, and may miss important evolutionary events. However, during the Cretaceous, siliceous sponges formed an important component of the chalk seas, and their silica often reprecipitated during burial to form flint. Flint nodules are most common in shallow water chalk, deposited in less then 100 m of water. Most modern hexactinellids, or glass sponges, are found in deep water, between 200 and 600 m, on the continental slope. They have also been dredged from abyssal depths. This suggests that the major move into deep water occurred in the Cenozoic.

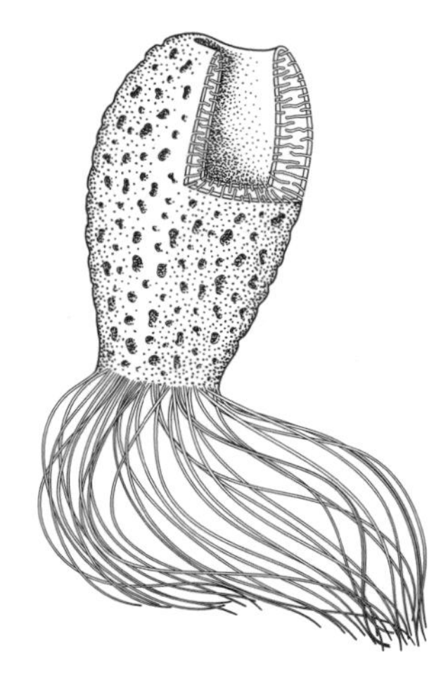

Fig. 3.3 A modern glass sponge found in deep water.

In the modern oceans, the combination of siliceous plankton and deep water siliceous sponges means that almost all biogenic silica is preserved in deep water sediments. In contrast to the Cambrian, the shelves are relatively starved of opaline silica, emphasizing the importance of evolution on major biogeochemical cycles.

Siphonia

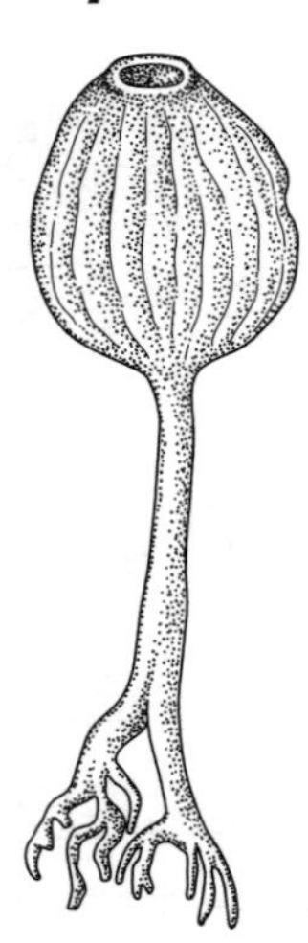

Demosponge

Cretaceous–Cenozoic
A tulip-shaped sponge with a bulbous cup developed on a long stalk, held firmly to the substrate by a root system (height approximately 10 cm). Commonly found within flint nodules in chalk. The most commonly preserved elements of this sponge are irregularly shaped, siliceous spicules. There is a leucon grade of organization, with a series of canals within the thick wall of the tulip head that would have accommodated a complicated series of filtering cells.

Rhaphidonema

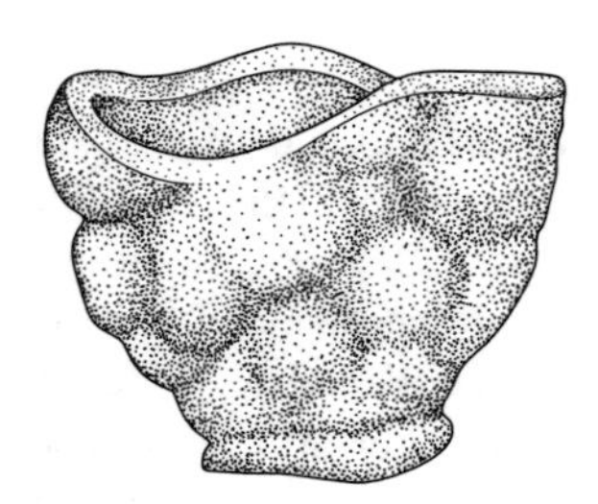

Calcarea

Triassic–Cretaceous
An irregularly shaped, roughly vase-shaped sponge composed of simple, calcarous spicules (diameter approximately 10 cm). A poorly developed internal canal system suggests a leucon grade of organization. The skeleton was robust and survives well in the fossil record. Specimens are often found in great abundance, suggesting that, in common with modern sponges, this species was gregarious and tended to live in large groups on appropriate, high energy substrates.

Glossary

Ascon – simplest grade of sponge organization: the animal composes a single cup with perforated walls.
Flagella – whip-like appendage on a cell, allowing it to move, or to generate currents.
Leucon – the most complicated grade of sponge organization: a series of feeding chambers are linked by canals to a common paragaster.
Osculum – aperture of paragaster.
Ostia – small pores in the sponge wall through which water enters the animal.
Paragaster – central chamber of sponge.
Sycon – intermediate grade of sponge organization: several feeding chambers open from a single paragaster.

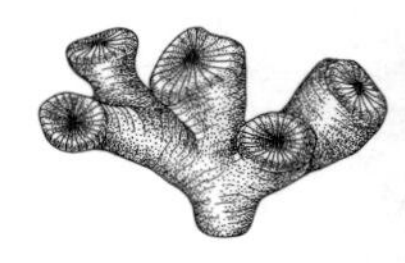

4 Corals

- Corals are amongst the most primitive multicellular animals.
- Corals are part of the phylum Cnidaria, along with jellyfish and sea anemones.
- Three diverse groups of corals arose independently from soft-bodied ancestors: the Tabulata, Rugosa, and Scleractinia.
- Rugose and tabulate corals were important members of the Palaeozoic community and major contributors to their reefs.
- Scleractinian corals are amongst the most important modern reef-builders.

Introduction

The cnidarians (pronounced "nidarians") are a phylum that includes jellyfish, sea anemones, and corals. Cnidarians evolved in the Precambrian, and are amongst the earliest multicellular animals to be found in the fossil record. They are simple metazoans, with a primitive grade of organization.

Although their bodies are organized into tissues, they lack organs. Their body plan is usually radial, although this has been modified in some groups. The cup-shaped body is composed of two layers of cells with a supportive filling. Food is digested within the cup, whose only aperture is usually surrounded by tentacles. Stinging cells characterize the tentacles and outer cell wall, and are used for defense or to catch prey.

Cnidarians have an unusual life cycle (Fig. 4.1), with a floating and a sedentary phase. A swimming larva settles on the sea bed and forms a polyp. This can bud to form asexual clones that may stay attached to the parent or break away. It can also form a sexual phase that floats in the plankton, aiding dispersal of the species. This sexual phase then produces larvae that settle. Jellyfish are the sexual phase of one such group, and sea anemones the benthic stage of another. The duration of each element of the life cycle varies, with some groups being long-lived on the sea bed and others in the plankton. Corals have abandoned the planktonic stage altogether and reproduce sexually from the sea floor.

The asexual reproduction of corals has important consequences for their lifespan. Each polyp is a clone, and shares a single genotype. This is true for all of the polyps living at one time, and also for all of the polyps that build the colony over time. In addition, one coral colony can live indefinitely, even though the individual polyps may die off. Corals in the Carribbean have been shown to be several thousands of years old and may have colonized the sea floor as it was flooded by the postglacial rise in sea level.

Geologically, the most important class of cnidarians is the Anthozoa, which includes the three orders of skeleton-building corals: order Tabulata, order Rugosa, and order Scleractinia.

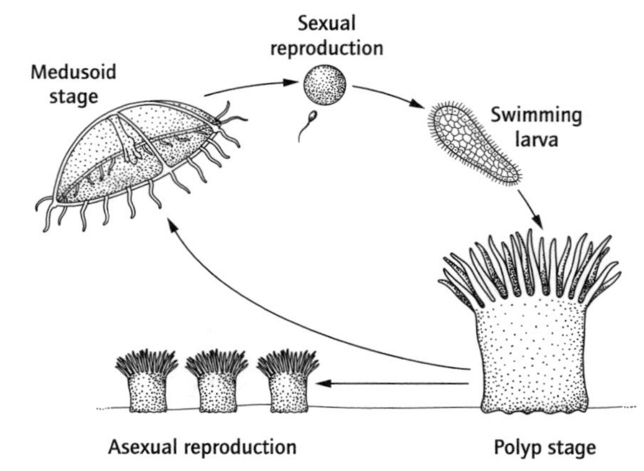

Fig. 4.1 Alternation of generations in cnidarians. Most cnidarians alternate between a planktonic sexual stage and a sessile asexual stage, which can reproduce by budding or cloning.

Morphology and evolution

The class Anthozoa includes soft-bodied sea anemones and corals that build calcareous skeletons. This ability to biomineralize arose at least four times within the group, once for each of the major skeletal orders, and once for an unusual Silurian coral called *Kilbuchophyllia* (Fig. 4.2). Coral skeletons can be made from aragonite or calcite ($CaCO_3$). Most tabulate and rugose corals are built from the latter, most scleractinian skeletons from the former.

Tabulate and rugose corals evolved from soft-bodied anemone ancestors in the Ordovician period. They thrived in the Silurian, Devonian, and, following a late Devonian decline, in the Carboniferous. They were part of the Palaeozoic reef community, but lacked a holdfast and so did not form the framework of reefs as modern corals do. They became extinct in the end-Permian mass extinction. Scleractinian corals evolved in the Triassic and radiated throughout the Mesozoic. Though many genera became extinct during the end-Cretaceous mass extinction event, they have come to dominate Cenozoic and modern reefs.

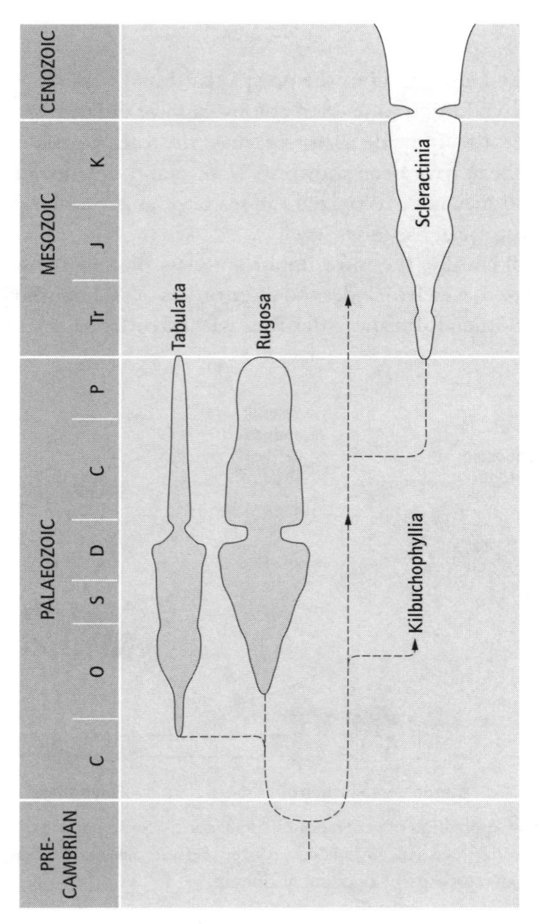

Fig. 4.2 The ranges and possible evolutionary relationships of the major groups of corals. Skeletons evolved separately several times within the group, and each coral order has a soft-bodied ancestor.

Many modern corals have a symbiotic relationship with an alga that lives within their tissue. These organisms, known as zooxanthellae, are protected by the coral and in return are "farmed" for nutrients. They change the internal chemistry of the animal, making aragonite secretion easier. These hermatypic corals are typically colonial. They must live within the photic zone so that their zooxanthellae can photosynthesize. They require clear water less than 30 m deep, and thrive in sea temperatures between 23 and 29°C. Ahermatypic corals live without an algal symbiont. They are typically solitary corals inhabiting deeper water. Some rugose and tabulate corals may have had zooxanthellae, but no direct tests for this hypothesis are available.

Corals grow a calcareous cup (corallite) surrounding the lower part of their soft tissues (polyp) (Fig. 4.3). As the coral grows, or buds, more mineral is secreted and the structure grows. The whole growth history of a coral is preserved in its skeleton. Within the corallite, a variety of vertical or horizontal structures are built to support the polyp. The most important vertical structures are radial septa (singular septum). The most important horizontal structures are plate-like tabulae (singular tabula) and smaller, upwardly curved dissepiments. Colonial corals have varying degrees of contact between the soft tissues of adjacent polyps. Sometimes the boundary walls between polyps are perforated, sometimes they are lost completely. The whole coral colony is called the corallum, and it can form a range of shapes, from erect, branching forms to low domes.

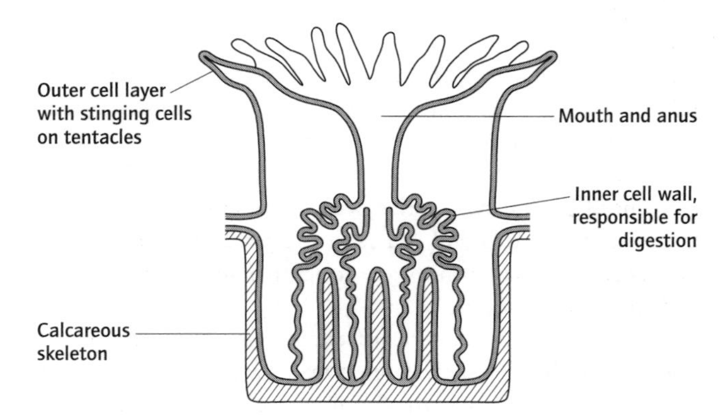

Fig. 4.3 Section through a generalized coral polyp to show the soft-part morphology. Medusoid stages are similar in morphology, but live "upside down" relative to this diagram, with the mouth hanging below a soft bell of tissue, and no skeleton.

Tabulate corals

Tabulate corals were always colonial, and the individual polyps tended to be small. At various times it has been suggested that they were not real corals, but recent work on their detailed skeletal structure shows that this is their true affinity. Preserved polyps from the tabulate genus *Favosites* have been discovered. Each polyp had 12 tentacles, and a similar overall appearance to the polyps of modern corals.

Tabulate corals first appear in Lower Ordovician rocks from North America, which was a low latitude continent at that time. They diversified rapidly in the Ordovician and quickly spread worldwide. A rapid radiation was followed by extreme decline in the end-Ordovician mass extinction. They recovered from this to reach a diversity peak in the middle Devonian, but their recovery from the late Devonian extinctions was restricted and they survived with limited diversity until the end-Permian extinction.

Large tabulate corals tend to be associated with Lower Palaeozoic reefs and small ones with deeper water facies. In particular, tabulates characterize reefs built by stromatoporoids (Chapter 3), which seem to have created many niches for them to occupy. This group of sponges became extinct in the late Devonian and its absence may explain the failure of tabulate corals to return to their previous diversity after this time.

Tabulate skeletons are made of calcite and tend to be very solid in form. The most diagnostic elements of the tabulate coral are the structures developed within the corallite (Fig. 4.4). These are dominated by the horizontal tabulae and dissepiments. Septa are short or absent. In some more advanced tabulates, the outer walls of the corallite can be thinned, or replaced completely by a marginal zone, shared between polyps, and filled with a boxy framework of internal struts, known as coenenchyme. Where corallites do have walls, they are usually perforated by mural pores, which would have allowed direct connections between the soft tissues of adjacent polyps. These adaptations indicate that there was a high degree of interconnection between the individuals making up a tabulate colony.

The overall shape of the colony was controlled partly by the substrate and partly by the pattern of addition of new corallites. These were either added around the edge of the colony, a pattern known as peripheral growth, or in between corallites, a pattern known as medial growth. Some species typically grew in only one pattern, while others could vary their pattern depending on the environmental conditions they experienced. Peripheral growth formed flat (tabulate) or low, dome-shaped colonies. Medial growth formed higher domes or nodular colonies. In a well-developed reef system, colonies in deeper water generally showed peripheral growth. Corals on the reef margin were dominated by medial growth and corals in the core of the reef showed both strategies, leading to a diverse array of colony shapes.

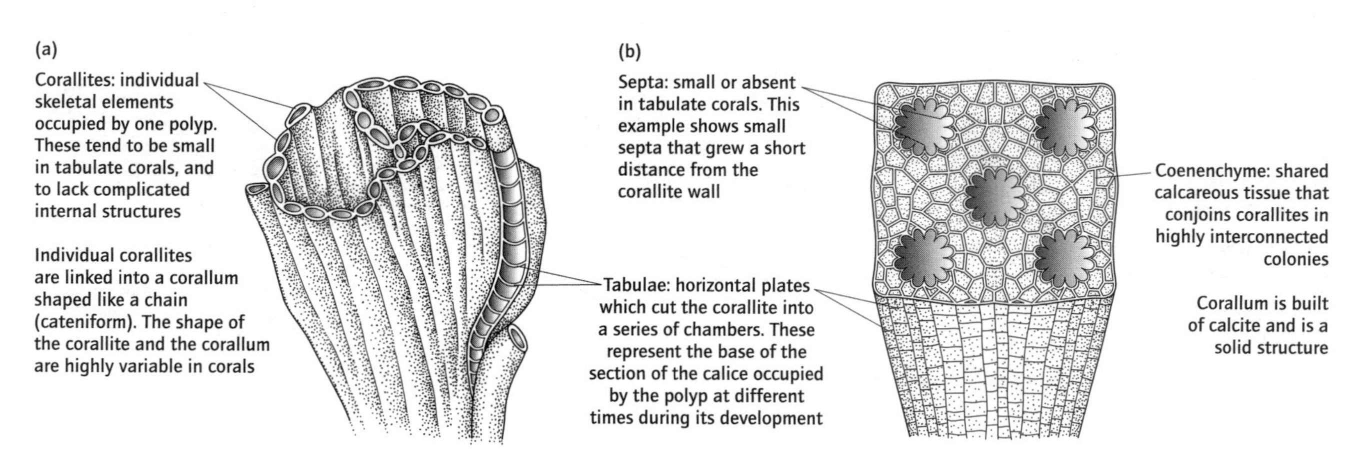

Fig. 4.4 Main features of the hard-part morphology of tabulate corals: (a) *Halysites*, and (b) *Heliolites*. Corallites of both of these genera are between 2 and 6 mm in diameter.

Rugose corals

Rugose corals first appear in the geological record in Middle Ordovician rocks from North America. They diversified more slowly than tabulate corals, but their patterns of evolution are similar. They were important members of Palaeozoic reef communities, but their diversity declined during the end-Devonian extinction. At this time, and in the earlier extinction event at the end of the Ordovician, solitary corals and generalist colonies were more likely to survive than highly specialized colonial forms. During the Carboniferous, rugose corals regained some diversity and were the more common of the two orders. Throughout their evolution, it has been suggested that there was a trend towards a more integrated colony form, with more contact between polyps of the same colony. All rugose corals became extinct in the end-Permian mass extinction event.

Rugose skeletons are almost always composed of calcite, though a few late forms may have secreted aragonite corallites. They have a solid structure similar to tabulate corals, though the internal elements of their corallites are distinctively different (Fig. 4.5). These are dominated by septa, vertical plates organized in a radial pattern. Tabulae and dissepiments are also common. In the center of the corallite there is often a central structure, which was produced by the modification of a variety of other internal structures. A coenenchyme, or region of shared tissue between corallites, is sometimes developed in colonial species.

Solitary corals are typically horn shaped; they reclined on soft sediment and grew upwards with time, so bending their corallite. Colonial corals tend to be dome shaped, with a range of corallite shapes, often defined by their proximity to one another. Corallites in direct contact tended to become polygonal in shape, whereas more isolated corallites kept a circular cross-sectional form. Within colonial rugose corals there is often evidence of the colony having fallen over and then regrown. This emphasizes the lack of holdfasts in the group, which precluded them from forming the framework of the reefs they inhabited. This lack of a holdfast also precluded rugose and tabulate corals from living in the high energy environments characteristic of the reef-front of a modern reef, which is exposed to breaking waves. Instead they were confined to the deeper, quieter slopes of the fore-reef, or the still lagoons of the back-reef area.

The overall shape of colonial corals is partly determined by their species, and partly by their environment. This is a useful property for interpreting mode of life, but makes taxonomy and classification very difficult. A tall colony can be interpreted as one that lived in an area of high sedimentation rate, but it need not be related to another colony with a similar form.

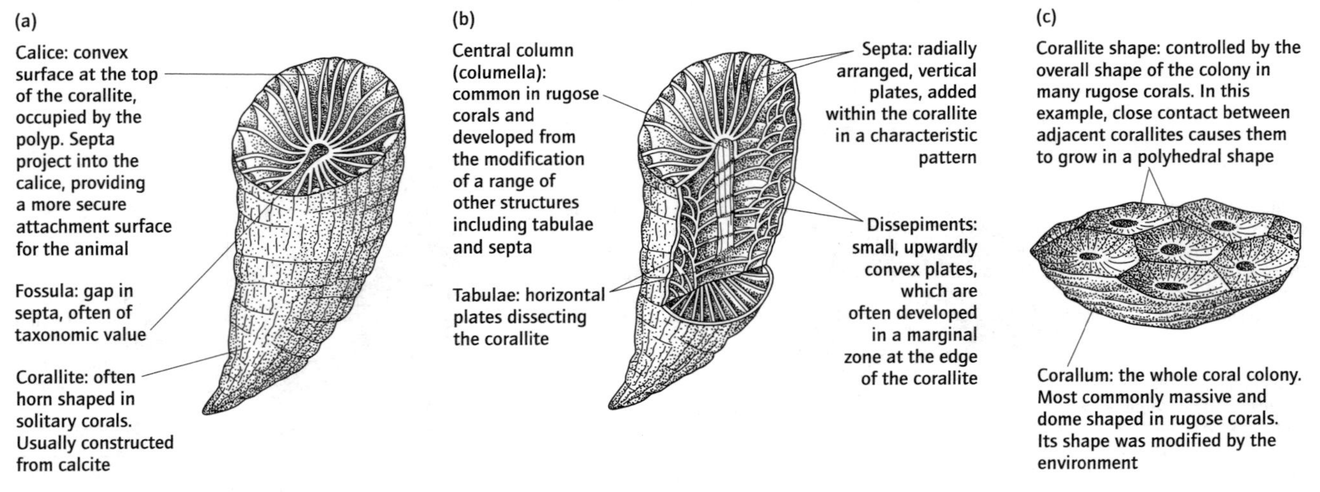

Fig. 4.5 Major features of the hard-part morphology of rugose corals: (a,b) a generalized solitary coral, and (c) a generalized colonial coral.

Rugose septa

In common with all corals, rugose corallites grew sequentially so that the whole history of their growth and development is preserved within the skeleton. This makes it possible to trace the development of the septa, which are added in a pattern characteristic of the group (Fig. 4.6).

In the early growth stages, the corallite had two opposing septa, known as the cardinal and countercardinal septa, respectively. Next, two alar septa were inserted on either side of the cardinal septum, and then two more septa on either side of the countercardinal, the counterlaterals. These four new septa defined four sections of the corallite, into which new suites of septa were inserted. This pattern of fourfold insertion is characteristic of the rugosa. It tends to leave gaps, most often in the region of the cardinal septum. These gaps, or fossulae (singular fossula), are taxonomically important.

The addition of septa in four areas of the corallite means that rugose corals secondarily lost their radial symmetry. Any functional reason for this is unclear, although it may have persisted into the soft tissues as they would have been in direct contact with the septa, and the shape of one should mirror the shape of the other.

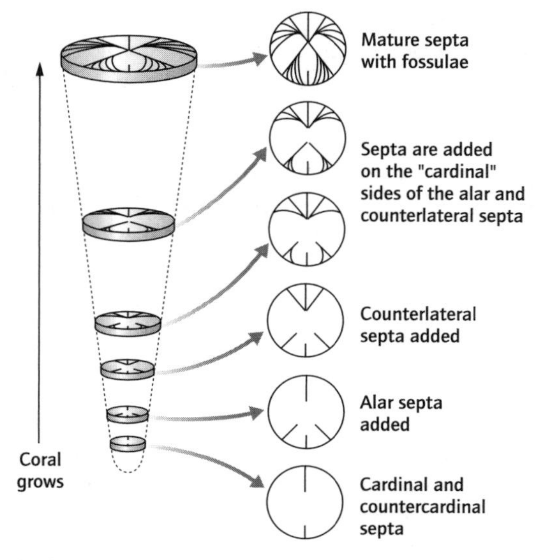

Fig. 4.6 The sequential development of septa within a rugose coral.

Lower Palaeozoic reefs

Reefs are rare in modern oceans, occupying around 0.2% of the sea floor. However, they have been much more common at some periods in the geological past, when the climate was warmer, sea levels were high, and large areas of shallow water occurred in temperate and tropical latitudes. The most important expansion of reefs seems to have occurred in the middle and late Devonian. They were very different to modern reefs, with different framework-builders and different groups of organisms filling the diverse range of niches available within the reef.

Devonian reefs were dominated by stromatoporoid sponges (Chapter 3). Tabulate corals growing in association with these sponges helped to stabilize the reef, as did rugose corals and calcified bacteria. However, the lack of holdfasts in the two coral groups impeded stable reef formation. Stromatoporoids lacked photosynthetic symbionts, and controversy exists about whether either group of coral had them. It seems likely that these reefs depended on filter-feeding more than algal farms for survival, and hence occupied areas of nutrient-rich waters. These reefs could be extremely large. The Barrier reef belt of the Canning Basin in Australia is 350 km long and does not seem to have been exceptional. Early cementation made the reef very strong, and blocks up to 100 m in diameter could break from the reef front and fall intact onto the fore-reef. There is good evidence that these reefs grew up to sea level and adapted to the physical demands of living in waves by developing a series of channels and grooves in the reef through which water could move.

In addition to corals, sponges, and bacteria, Devonian reefs were colonized by a wide diversity of brachiopods, trilobites, byrozoans, and more problematic organisms. Predation and boring were relatively restricted compared to modern reefs.

Scleractinian corals and reefs

Scleractinian corals evolved from soft-bodied ancestors in the Middle Triassic period. By late Triassic times they had begun to form small patch reefs, and their importance as reef-builders has been continuous since then. They are facilitated in this role by the following morphological adaptations: they have a basal plate which acts as a holdfast, they build porous skeletons of aragonite (which are more readily secreted than the massive calcite skeletons of more primitive corals), they are able to add material to the outside of the corallite to cement it to a firm substrate or adjacent colony, and they can live in a symbiotic association with photosynthetic zooxanthellae.

Solitary scleractinian corals evolved in the early Jurassic, and inhabited deeper water. They became diverse and important late in the Cretaceous. Many genera of both types of coral disappeared as a result of the end-Cretaceous extinction event. Generalists, with a wide ecological range, appear to have had a better chance of survival than specialist forms, though there is no apparent difference in survival rates between those with zooxanthellae and those without.

Internally, the corallite of a scleractinian coral tends to be dominated by septa (Fig. 4.7). Dissepiments and central columnar structures may also be developed. Septa are added in cycles of 6, 12, or 24, each with a regular spacing. The walls of the polyp hang over the edge of the calice, explaining how aragonite can be secreted on the outer surface of the corallite. Colonial scleractinian corals have well-integrated soft tissues and often lack corallite walls. These are replaced by a shared zone of perforated aragonite, similar to the coenenchyme of rugose and tabulate corals, but known as the coenosteum.

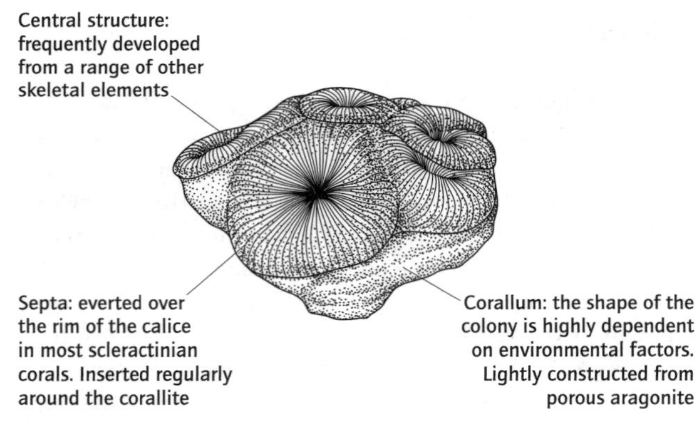

Fig. 4.7 Major elements of the hard-part morphology of scleractinian corals. This example is *Confusastraea*.

Scleractinian corals are amongst the most important reef-builders of the Mesozoic and Cenozoic. Reefs commonly develop as fringes around small islands. If sea levels rise, corals can often keep pace with the rate of change, building upwards and outwards towards the high energy zone of wave action. In doing so they migrate away from the shoreline over time, and may eventually be the only remnant of a sunken island, forming a ring-shaped atoll around the drowned land.

Reef-forming corals are amongst the small number of organisms capable of modifying their environment, changing the topography of the sea bed in such a way as to promote their own survival. Incidentally, this increases local biodiversity by generating a range of spatially distinct niches. For example, structures such as the Great Barrier Reef are of enormous extent and geological importance.

These diverse ecosystems exist within low nutrient regions of the oceans, such as around mid-ocean islands. In earlier geological periods these appear to have been regions of much lower diversity.

Scleractinian corals reproduce without undergoing a planktonic phase. This leads to problems of dispersal, as the gametes cannot travel far from their parents before settling. This in turn has led to a pronounced provincialism within modern corals, with distinctive modern Indo-Pacific and Caribbean provinces developed. Diversity is much higher in the Indo-Pacific region (700 species, compared to 62 in the Caribbean), suggesting that this region was a refuge for corals during the low sea level of the major Pleistocene glaciations, and the location from which their subsequent radiation has occurred.

Corals as climatic indicators

Corals have a long lifespan and record events that happened during their whole life in the incremental growth of their skeletons. Solitary rugose corals, such as the Silurian genus *Kodonophyllum*, narrow and thicken along their length (Fig. 4.8). Wide patches represent times when the polyp was thriving. Narrow zones indicate times when the polyp was under stress and lost body mass, contracting towards the center of its calice, and only adding skeletal material to this central portion.

Corals with zooxanthellae are confined to a narrow range of environmental conditions. The presence of colonial corals in a rock sequence, especially scleractinian corals, is an indication that a region was within 30° of the equator when the rock was deposited. It is only within this region that the water temperature is between 23 and 29°C, the range in which reef corals thrive. The presence of these corals also indicates that deposition occurred in clear, shallow, nutrient-poor water (Fig. 4.9). Such a set of assumptions is less secure for rugose and tabulate corals as it is not known whether these forms had photosynthetic zooxanthellae. However, by using independent environmental indicators, studies on Lower Palaeozoic reefs, such as the Wenlock Limestone of Shropshire, confirm that they developed in shallow water depths and at low latitudes.

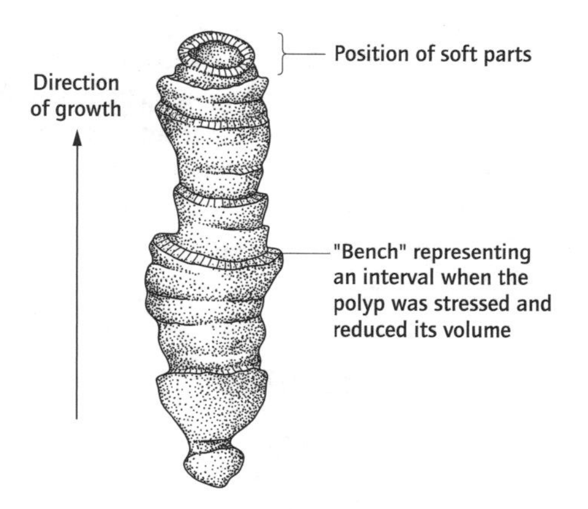

Fig. 4.8 *Kodonophyllum*, a Silurian solitary rugose coral, showing periods of restricted growth and periods when the polyp was larger, representing favorable conditions.

Reef corals of Pleistocene age or younger can be used to generate a much more detailed record of palaeoclimatic change, preserved within a single colony, or in a series of colonies of overlapping ages. As the colony adds calcium carbonate (in the form of aragonite) to its skeleton, it removes carbon and oxygen from seawater. Each of these elements is stable with more than one atomic weight, and is thus said to have several stable isotopes. Heavier and lighter isotopes of the same element are fractionated in the oceans by a range of climatic events and these are recorded in the skeletons of growing corals. For instance, water evaporating from the oceans is enriched in the lighter isotope of oxygen. At times in the past when this water was locked up in ice on land, the oceans become isotopically heavy in oxygen, and this is a useful palaeoclimatic signal. The ratios of stable isotopes of these elements in water, recorded as $\delta^{13}C$ and $\delta^{18}O$, are hence good proxies for determining palaeoclimate. In addition, corals incorporate traces of organic material into their skeleton, including humic acids from river outflow, derived from biological weathering. These show up in sections cut through the coral as thin bands that fluoresce in UV light. Bands within the coral where humic acid is abundant indicate that the colony was growing in periods of high runoff from nearby rivers.

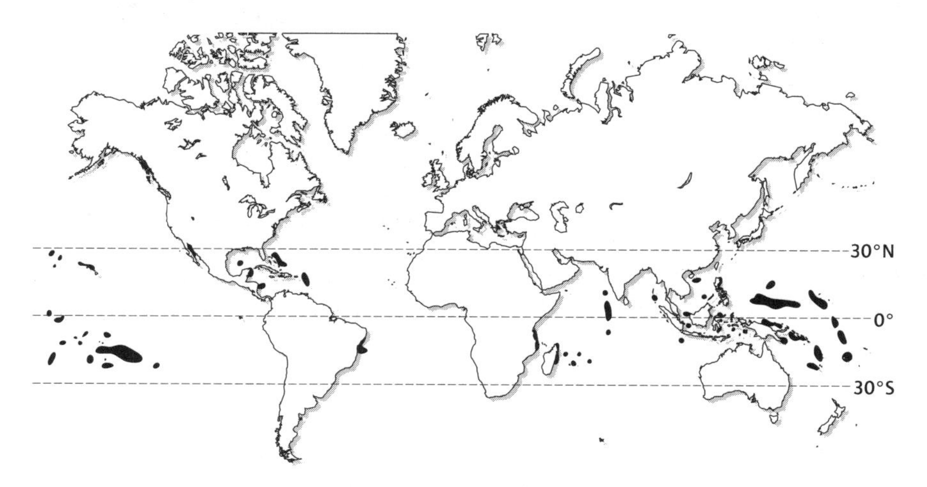

Fig. 4.9 Map of the modern world showing the current distribution of coral reefs within 30° of the equator. Note the abundance of open ocean sites that are areas of low nutrient availability.

Favosites

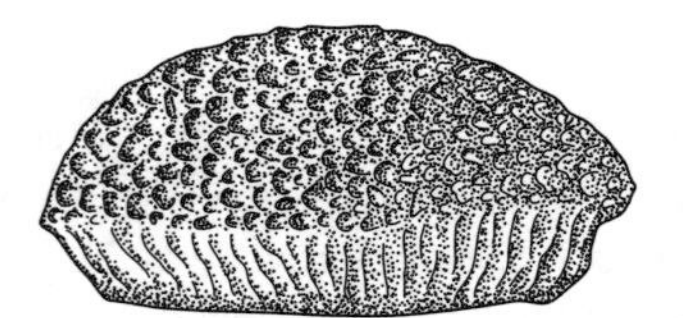

Tabulate coral

Upper Ordovician–Middle Devonian

A small (typically 3–5 cm diameter) colonial coral. Corallites are polygonal and cerioid, that is, each corallite retains its wall. Each is about 2 mm in diameter. Within the corallites short septal spines are developed. Tabulae are abundant and regular in their spacing. Mural pores connect the corallites. They are common in Lower Palaeozoic reefs.

A Silurian *Favosites* colony from Quebec preserves the calcified remain of polyps within their calices. This unusual soft-part preservation helps to confirm the position of tabulates as true corals.

Palaeosmilia

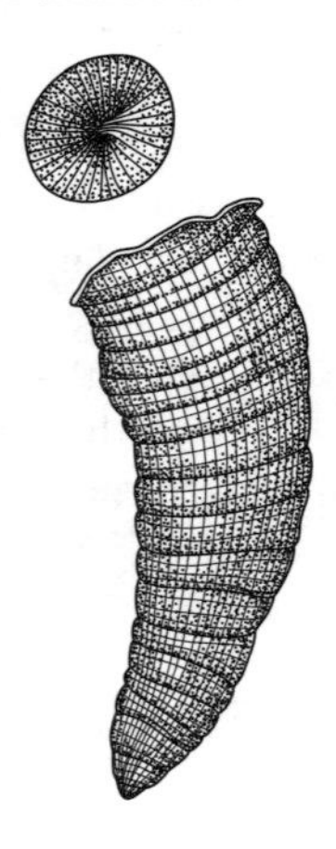

Rugose coral

Carboniferous

A large solitary coral, with a corallite often reaching 3 cm in diameter and up to 20 cm in length. Septa are numerous and may fuse at the center of the corallite. A wide band of dissepiments is developed in a marginal zone.

Halysites

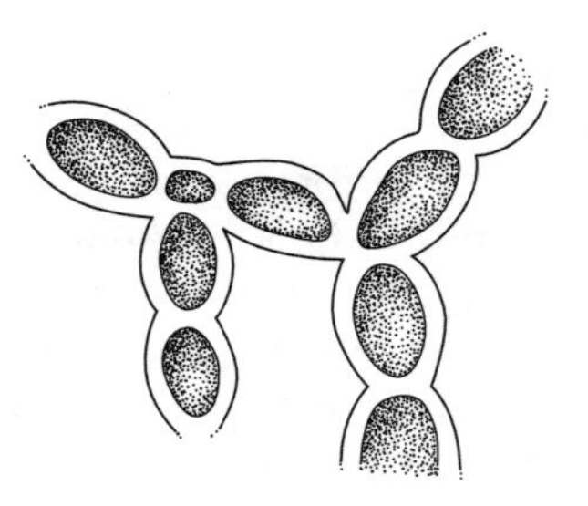

Tabulate coral

Middle Ordovician–Silurian

A colonial coral with a distinctive chain-shaped corallum. This cateniform shape is rare. Each corallite is connected to the next by a coenenchyme composed of a single tube cut horizontally by many small plates. Corallites are usually 2–6 mm across.

The chain form of this coral had several advantages, especially in areas of high sedimentation rate. The holes between the polyps allowed them space to spread out for feeding, and also to dispose of unwanted sediment. They were able to colonize large areas of the sea bed very rapidly early in their development, and then to grow upwards to keep up with sediment influx.

Isastraea

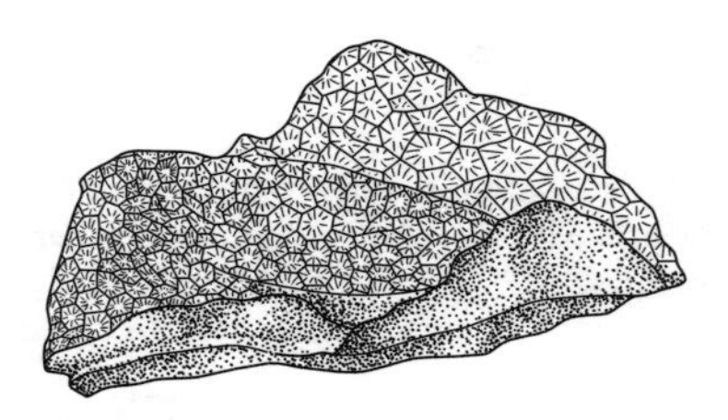

Scleractinian coral

Middle Jurassic–Cretaceous

A massive, colonial coral with a cerioid corallum. This was a reef-dwelling coral, with a strong construction analogous to hexagonal close packing. It would have thrived in high energy zones of low sedimentation rate. Feeding efficiency may have been compromised by the lack of space between polyps. Individual polyps were 4–7 mm in diameter.

Lithostrotion

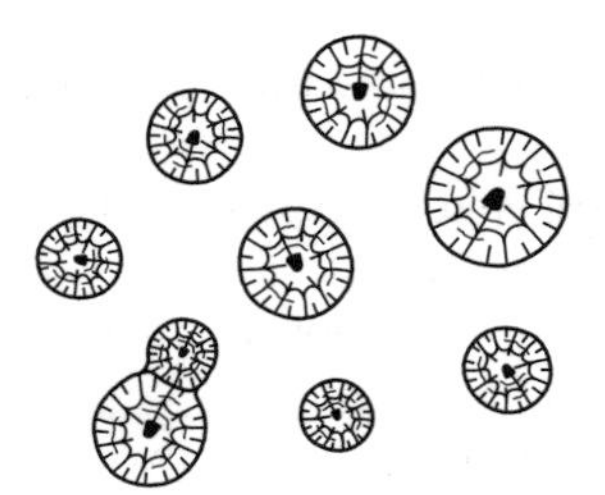

Rugose coral

Carboniferous
Corallites are typically less than 1 cm in diameter. A small central bar-like columella is developed and there is a pronounced edge zone composed of dissepiments. However, colonies can exceed 1 m across.

This coral genus shows a great deal of environmentally determined variation. Corallites within the colony may be separated from one another, a form known as fasciculate. However, they may also be in close contact with one another with the shape of the corallites modified accordingly.

Dibunophyllum

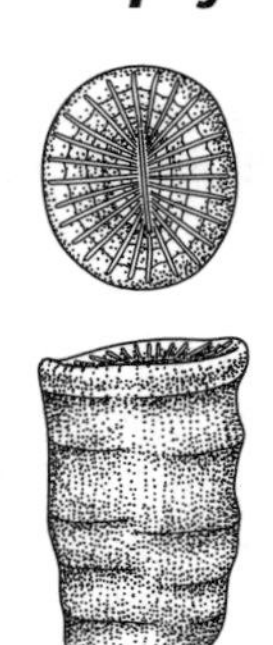

Rugose coral

Carboniferous
A cylindrical colony with a diameter of 3–4 cm and a length of around 10–15 cm. The most distinctive feature of this genus is a prominent central structure that looks like a spider's web in plan view. There are many septa and a narrow edge region is filled with dissepiments.

Thecosmilia

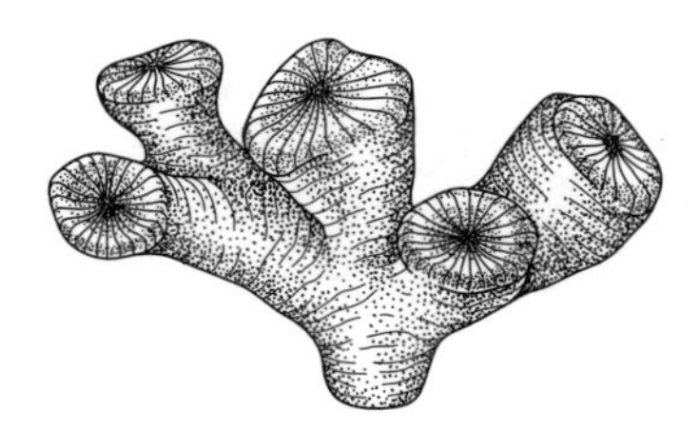

Scleractinian coral

Middle Jurassic–Cretaceous
A colonial form characteristically composed of a few large corallites, perhaps 10 in a colony, each with a diameter of 2–3 cm. The corallite form is similar to that of *Montlivaltia* (right). The corallum shape is fasciculate, and the corallites usually branch one from another.

Montlivaltia

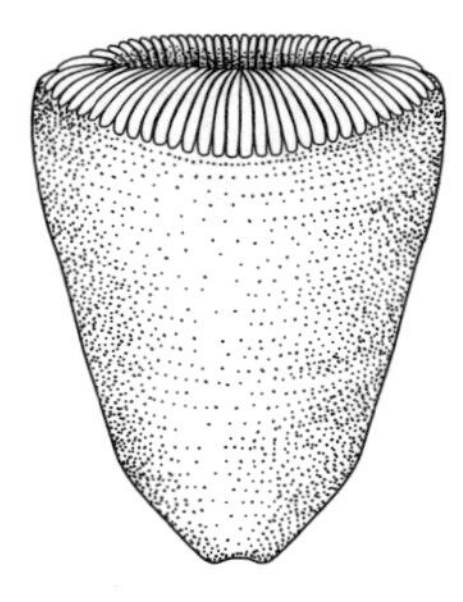

Scleractinian coral

Middle Jurassic–Cretaceous
This is a squat, solitary coral with a diameter of 2–3 cm and a length of 3–5 cm. The abundant septa project a significant distance above the shallow calice, and are abundant. There are many dissepiments but no columella. This was a reef coral, which despite its solitary habit probably had zooxanthellae.

Glossary

Ahermatypic – corals lacking symbiotic algae.
Calice – cup-like depression on top of a corallite, in which a polyp lived.
Cateniform – coral colony shaped like a chain.
Cerioid – coral colony where individual corallites retain their walls, but are in contact with one another.
Coenenchyme – shared hard tissue between the corallites of rugose and tabulate corals.
Coenosteum – shared hard tissue between the corallites of scleractinian corals.
Columella – central or axial structure within a corallite, formed from a range of other internal features including tabulae and septa.
Corallite – hard exoskeleton of a single coral polyp.
Corallum – hard parts of a whole coral, solitary or colonial.
Dissepiments – small, outwardly convex plates developed within a corallite.
Fasciculate – coral colony where individual corallites are not in contact with one another.
Hermatypic – colony with photosynthetic zooxanthellae, usually colonial and reef-forming.
Mural pores – connecting hole between adjacent corallites within a colonial coral.
Polyp – single coral animal.
Septa – vertical plates, with a broadly radial arrangement within a corallite.
Tabulae – horizontal plates within a corallite.
Zooxanthellae – photosynthetic, symbiotic algae living within the tentacles and upper surface of hermatypic corals.

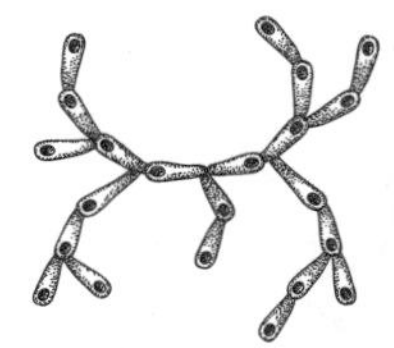

5 Bryozoans

- Bryozoans were amongst the most diverse organisms of the Palaeozoic.
- They are all colonial and almost all marine
- They live as benthic filter-feeders, usually attached, sometimes free lying, occasionally mobile.
- Individuals are typically less than 1 mm in size, but colonies can approach 1 m in diameter and contain over one million individuals.

Introduction

Bryozoans are filter-feeding, colonial animals that have formed a significant part of the marine benthos since the Ordovician. They are sometimes known as moss animals and they superficially resemble plants rather than animals. They form a phylum of their own, probably most closely related to brachiopods. Around 20,000 species are recognized, most of these from the fossil record.

Bryozoan zooids are tiny and feed via a lophophore – an array of tentacles that extract food particles from water. Food is moved through a U-shaped gut to an anus located just outside the ring of tentacles making up the lophophore. Respiration occurs by diffusion as the animals are so small, and they have no circulation system or gills. Bryozoans have a relatively sophisticated nerve system and a complicated series of muscles helping them to move in and out of their skeleton. Male and female characteristics can occur within the same zooid or the same colony. In addition, most bryozoan colonies have a range of specialized zooids that clean or protect the colony, or are dedicated to breeding. The strangest of these specialized zooids are the avicularia. These have highly adapted, toughened areas that are usually used in defense or cleaning, but in some cases can be used as stilts on which the colony "walks".

All of the individuals in a bryozoan colony are genetically identical, regardless of their degree of specialization. They grow by budding from a single individual, called the ancestrula. The colony is built of gelatinous or fibrous protein, aragonite, or calcite, or a mixture of these materials. The shape of the colony reflects the need to feed efficiently and the demands of substrate colonization. A restricted range of shapes have evolved many times, a process known as iterative evolution. These include closely packed mats, elongate shapes, runner forms, erect tubes, disc shapes, and upright fans.

Bryozoans are common in most shallow marine environments. They can also colonize deeper water and a few live in fresh water. In reef environments they can form important sediment baffles that are full of small cavities, providing cryptic niches for a wide diversity of organisms. Bryozoans are rock-formers, sometimes to a significant extent, for example during the Carboniferous period. They are able to colonize most substrates, but prefer hard surfaces. In modern seas, bryozoans are the most important carbonate producers on the southern Australian and New Zealand shelves.

Marine bryozoans are divided into two classes, Stenolaemata and Gymnolaemata. The stenolaemates are highly calcified forms, with zooids living in tubes that grow throughout the life of the colony. Gymnolaemates are generally less heavily mineralized, and their zooids grow elegant boxes of fixed size. The stenolaemates were dominant in the Palaeozoic and for most of the Mesozoic, until the late Cretaceous. Since then one group of gymnolaemates, the cheilostomes, have dominated bryozoan faunas.

Bryozoan morphology

The hard parts of an individual zooid are called the zooecium, and the skeletal colony the zoarium (Fig. 5.1). The zooecia of stenolaemates are tube shaped and are often studied in thin section. In branching colonies, the mature parts of the zooecia usually grow at a high angle to the axis of colony growth. Zooecia may change in shape as they grow. They share common skeletal walls with adjacent zooids. In cross-section these tubes can be identified along with the shared hard tissue between.

In gymnolaemates the box-like zooecia usually develop nearly parallel to the growth axis of the colony. The upper surface of the zooecium may be protected by spines or a calcite shield to deter predation. In addition, there is a lid or operculum that fits over the aperture from which the zooids emerge. Different types of aperture or zooecia characterize different specialized zooids. More zooecia are added as the colony clones new individuals, and each one quickly reaches a finite size where growth ceases. As a result, the colony builds outwards in a modular fashion, adding building blocks of a standard size.

The shape of the colony, or zoarium, depends on the species and on the environment in which an ancestrula settles. Stick-like, fan-shaped, disc-shaped, and encrusting colony forms are common although they may be difficult to relate straightforwardly to taxonomy or ecology.

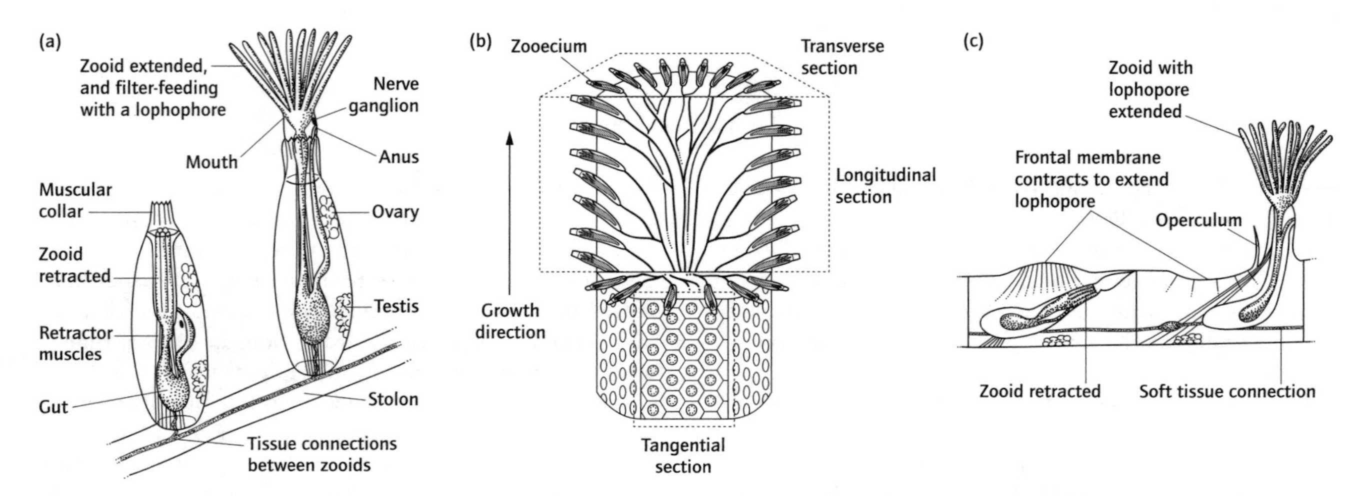

Fig. 5.1 Hard- and soft-part morphology of bryozoans: (a) ctenostome gymnolaemate zooids, (b) sections of a stenolaemate zoarium, and (c) gymnolaemate zooids.

Bryozoan ecology

Bryozoans have variable colonial integration, with zooids tending to become more closely integrated through evolutionary time. Specialized zooids, such as the avicularia, do not feed and are provided for by other members of the colony. The extremely complicated and precise colony forms of many bryozoans suggest a close integration of the colony, with overall control being exerted on the activities of individual animals building the skeleton.

The shape of the colony is often determined by the need to optimize the flow of water across the feeding zooids. These zooids act in concert to produce an even and optimal flow of water over the colony. Systematic variations in the orientation of zooid apertures, and the development of areas of the colony called maculae, exhalent "chimneys" into which these zooids send filtered water, are examples of this. In fan-shaped colonies, regular perforations occur in the fan, and zooids tend to feed only on one side of the structure. Water passes over the feeding surface and out through the perforations, producing an efficient flow pattern.

Encrusting colonies of bryozoans show a range of strategies for surviving in the competitive world of flat submarine surfaces. These colonies usually fall into one of three shape categories: sheets, spots, or runners. Sheet forms often have spines and raised margins to the colony, which impede rival growths. They tend to vent filtered water off the edge of the colony, at competitors. Runners cover a large area of substrate quickly, looking for refuges in inhospitable terrane. Spot colonies are short lived and depend on finding refuges that are inhabitable for a short time. They are sexually precocious and small.

Bryozoan evolution

There are five orders of stenolaemate bryozoans, and four of them evolved to their greatest diversity during the Palaeozoic. These include the well-known fan-shaped Fenestrata and the stony, usually stick-like, Trepostomata. These orders were severely affected by the end-Permian mass extinction event and all four were extinct by the end of the Triassic. The remaining order of stenolaemates, the Cyclostomata, were a minor component of Palaeozoic faunas but survived into the Mesozoic and radiated to a great abundance, especially during the Cretaceous. Many genera of cyclostomes became extinct at the end of the Cretaceous, but a few survive to the present day.

Gymnolaemate bryozoans are known from the Ordovician, but reached their acme in the Cenozoic. They are divided into two orders: the minor Ctenostomata, which have been a small element of bryozoan faunas throughout the Phanerozoic, and the hugely diverse order Cheilostomata. This group first appeared in the Jurassic and have come to dominate bryozoan assemblages (Fig. 5.2). Ctenostomes are entirely soft bodied and are usually preserved as borings on calcareous substrates. Cheilostomes are the most highly evolved bryozoans, with the most diverse and integrated colonies.

It is sometimes argued that cheilostomes outcompeted cyclostome bryozoans around the end of the Cretaceous. However, it appears that although both groups have similar ecological tolerances they have continued to co-occur in assemblages. The relative decline of one group cannot be attributed with certainty to the rise of the other.

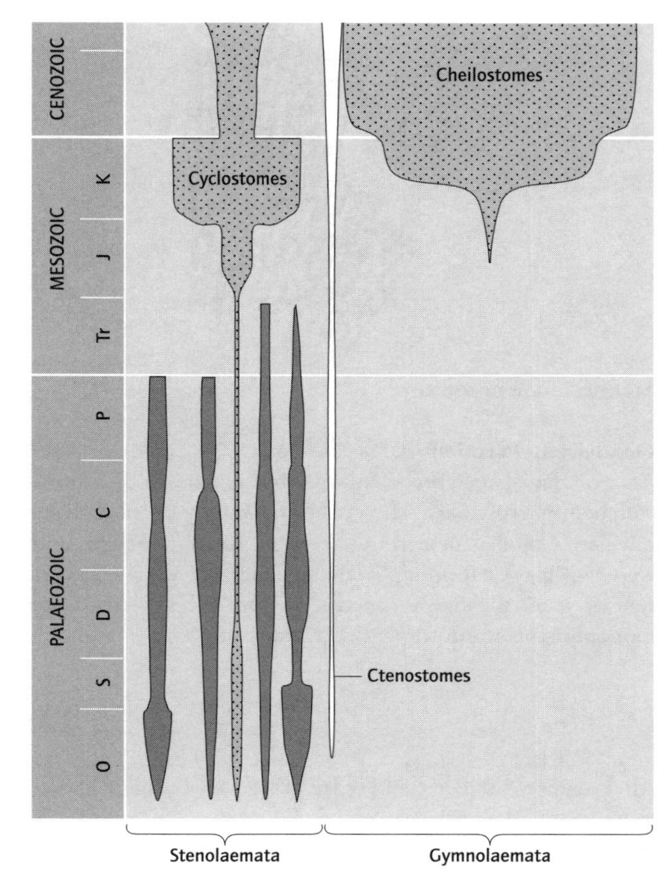

Fig. 5.2 Range and abundance of bryozoan classes. The dark shaded classes were the dominant Palaeozoic forms and are all stenolaemates. The dotted forms were the dominant Mesozoic and Cenozoic forms, the stenolaemate cyclostomes, and the gymnolaemate cheilostomes.

Bryozoans as environmental indicators

Bryozoans are potentially useful as environmental indicators, although their application to this topic is rarely straightforward. A major problem is comparing modern, cheilostome-dominated assemblages with older faunas. Rare bryozoan species are tolerant of most conditions, so statistical analyses of diversity or abundance tend to be applied. In post-Palaeozoic rocks and in the modern oceans, bryozoans are dominant members of shallow benthic communities in temperate latitudes, with normal salinity and low to moderate rates of sedimentation. Abundance in the modern day peaks at water depths of 40–90 m, and in sedimentation rates of less than 100 cm per thousand years. In common with most organisms, a high diversity usually represents environments close to the ideal range, while high abundances of a small number of species can typify more extreme conditions.

Individual shapes and sizes may also be used as environmental indicators in some cases. In general it has been observed that colony and zooid size decrease with increasing depth, although, paradoxically, zooids tend to be larger in cooler water. The shape of a bryozoan colony may be more or less favored in different environments. This means that environmental data may be buried in lists of the relative abundance of different morphological types. For example, encrusting forms are more common than erect shapes in shallow water in modern oceans, whilst rigid, erect colony shapes are most common from deep water assemblages. Free-living, mobile colonies are typical of sandy sea beds, where the bryozoan colony needs to be able to respond to a substrate that is itself mobile.

Fenestella

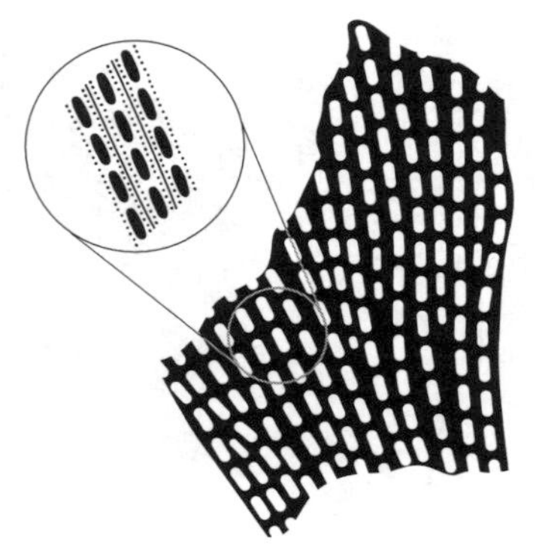

Stenolaemate bryozoan

Ordovician–Permian

An erect, fan-shaped bryozoan with the branches of the colony stiffened by cross bars. The skeleton is lightly to moderately calcified. Colonies ranged in size up to 20 cm in height, and were locally rock-forming. Each fan generated a one-way current of water through its apertures, from the side that bore lophophores towards the side that had none.

Stomatopora

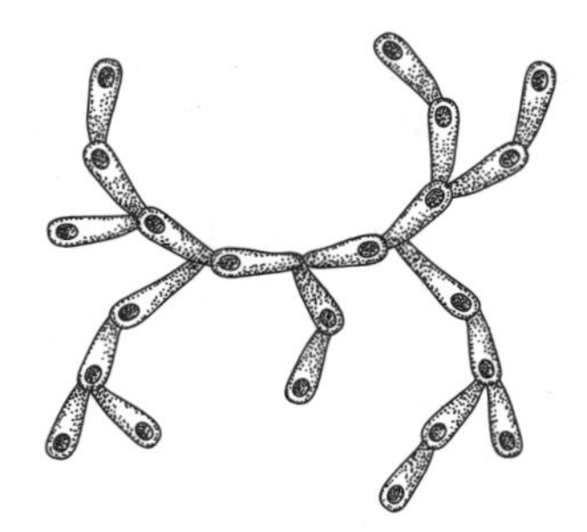

Cyclostome gymnolaemate bryozoan

Triassic–Recent

An encrusting bryozoan, evolved to a growth form that spreads the colony over a wide area with a low density of zooids. By dispersing zooids widely, it is more likely that the colony will survive following the local destruction (e.g., by predators) of zooids occupying one part of the substrate. As a member of a pioneer community, it would have encountered uncertain conditions, which might have been highly favorable or unfavorable.

Glossary

Ancestrula – founder zooid of the colony produced by metamorphosis of the settled larva.
Avicularia – specialized zooids that may defend the colony or, rarely, use their appendages as stilts on which the colony can "walk".
Operculum – hard "lid" that covers the aperture of the zooids of gymnolaemate bryozoans.
Zoarium – bryozoan colonial skeleton.
Zooecium – bryozoan zooidal skeleton.

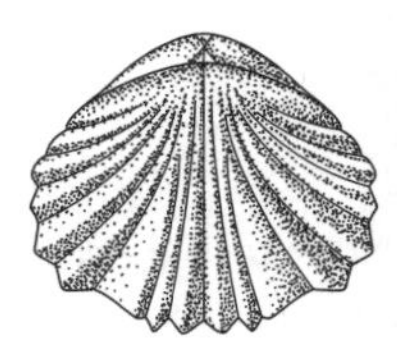

6 Brachiopods

- Brachiopods were the dominant shelly marine invertebrates of the Palaeozoic.
- They are classified into three subphyla: Linguliformea and Craniiformea ("inarticulated"), and Rhynchonelliformea ("articulated").
- They are exclusively marine filter-feeders.
- Brachiopod shell shape is sometimes indicative of substrate type.
- Communities of brachiopods can be used to study palaeoenvironments.

Introduction

Brachiopods were the most abundant and diverse marine invertebrates of the Palaeozoic. Brachiopods originated in the early Cambrian and diversified in the Ordovician. They dominated the shallow marine environment throughout the Palaeozoic. Although they survived the end-Permian mass extinction they declined through the Mesozoic so that of the 4,500 fossil genera known, only 120 exist today.

Brachiopods are superficially similar to bivalves; both have a hinged shell enclosing their soft tissues. This two-valved form arose independently in the bivalves and brachiopods and they are distinguished by their planes of symmetry. Brachiopod valves are symmetric about a medial plane bisecting the shell from the midpoint of the hinge to the center of the anterior edge. Valves of bivalves are usually mirror images of each other and the plane of symmetry runs between them, along the commissure.

Although brachiopod morphology seems relatively straightforward there are difficulties with their taxonomy because many species that look very similar have a very different internal organization. New techniques have allowed brachiopod classification to be reorganized on a line that more accurately reflects their family tree. The traditional division into two classes, Inarticulata and Articulata, has been superseded by a system separating brachiopods into three subphyla, the simplified characteristics of which are shown in Table 6.1. Linguliformea and Craniiformea may be considered to be inarticulated brachiopods and Rhynchonelliformea to be articulated brachiopods in the older classifications.

Most modern brachiopods live in marginal environments attached to the substrate and species tend to be morphologically similar. However, Palaeozoic brachiopods exploited a wider range of marine environments and their morphology was extremely diverse, ranging from erect coral-like forms to flattened saucer shapes. The morphology and community structure of brachiopods can be used in the interpretation of past environments.

Table 6.1 Brachiopod classification.

	Phylum Brachiopoda		
	Subphylum Linguliformea	**Subphylum Craniiformea**	**Subphylum Rhynchonelliformea**
Shell composition	Organophosphate	Calcareous	Calcareous
Hinge mechanism	Lacking teeth and sockets	Lacking teeth and sockets	Teeth and sockets present
Pedicle	Present	Reduced or absent	Present
Digestive tract	Gut with anus	Gut with anus	Gut without anus

Internal morphology

The internal anatomy of brachiopods can be divided simply into two sections. The pedicle, main internal organs, and muscles are packed together at the back of the shell whilst the lophophore dominates the mantle cavity in the central and anterior area (Fig. 6.1a). Food particles are trapped by the lophophore's cilia, passed back to the mouth and into the digestive tract. Waste products are emitted as small pellets from the anterior of the shell.

The pedicle is the primary method of attachment. Depending on the mode of life of the species it varies from being a thick muscular stalk to a series of thin threads. Rhynchonelliform brachiopods, which possess teeth and sockets, have two sets of muscles positioned close to the hinge area at the posterior of the shell (Fig. 6.1b). Adductor muscles, attached perpendicularly to the interiors of the dorsal and ventral valve, pull the two valves together when contracted and close the shell. Diductor muscles are attached obliquely to the ventral valve, just outside the adductors, and to the cardinal process of the dorsal valve. They pull on the dorsal hinge area when contracted and cause the shell to open. Brachiopods without a mechanical hinge have a more complex musculature that extends further into the shell and reduces shearing between the valves. Patterns of muscle scars preserved on the interior of the valves are species specific. In some brachiopods muscles are raised off the floor by platforms. Such structures are particularly useful in classification of certain rhynchonelliform orders, for example pentamerids.

The lophophore dominates the mantle cavity and is responsible for collecting suspended food from seawater. It generates currents to draw water into the shell, and removes food with sticky, fine tentacles. In brachiopods the lophophore is developed as a pair of tentacle-bearing arms. The overall shape varies from a simple horse-shoe to complicated folded configurations. It is an immovable, fixed structure held in place by the pressure within the brachiopod shell or supported by skeletal elements in the dorsal valve (Fig. 6.1b).

When a brachiopod feeds the valves open slightly and water is drawn in and discharged through separate chambers created by the lophophore. Currents are generated by small cilia that drive the water between the tentacles. Food particles (any organic particles, particularly phytoplankton) are transported to the lophophore ridge which has a brachial groove that conducts food to the mouth.

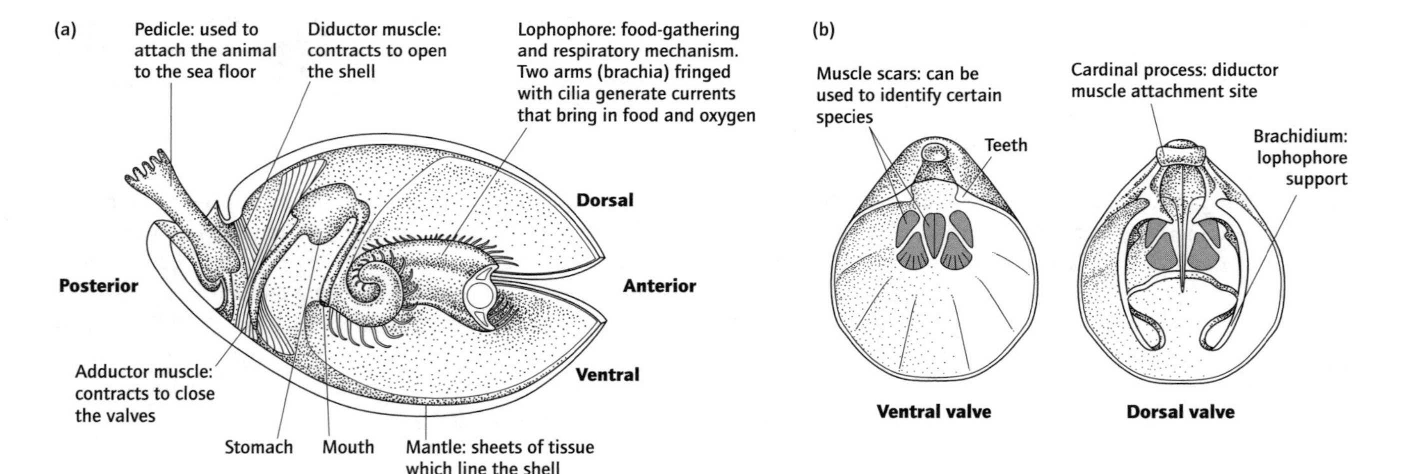

Fig. 6.1 Brachiopod internal morphology: (a) soft tissue, and (b) hard parts.

External morphology

Brachiopod shells grow by accretion. Mineralized material is secreted from the mantle lining the shell. Linguliforms are phosphatic, and craniiforms and rhynchonelliforms are calcareous. In the latter group the shell is multilayered and may have thin tubular structures perpendicular to the layering. These are called punctae and are a useful diagnostic feature of brachiopods in thin section. Key aspects of the brachiopod external morphology are: shell shape, shell sculpture, and the form of hinge area (Fig. 6.2).

Brachiopod shape is determined by the curvature of the valves. In order to accommodate the soft parts, at least one valve is always convex (has a rounded shape in cross-section). Both valves may be convex (biconvex) and the degree of roundness may be the same or unequal. Alternatively, one valve may be flat or concave (curves inwards) producing variations in the shell profile. Brachiopods commonly have an exterior surface texture. This may be in the form of ribs radiating from the beak, growth lines, or wrinkles.

The line of closure of the valves (commissure) may be straight or corrugated. It may also have a deep medial depression (sulcus) and a corresponding elevation (fold). The hinge area is very important in brachiopod classification. The hinge line may be straight (strophic) or curved (astrophic). The pointed extremity marking the start of valve growth is known as the beak and each valve has one. The area between the beak and the hinge line is known as the interarea. This may be flat or curved. In some brachiopods the beak is more prominent and curves over. In this case the posterior extremity is called the umbo.

The hole for the pedicle is called the pedicle foramen. It is sometimes closed by a single plate (deltidium) or plates (deltidial plates). Rather than a rounded hole, in some brachiopods the opening for the pedicle is more of a notch (delthyrium). This gap may be extended to the dorsal valve (notothyrium) to enlarge the opening.

In epifaunal brachiopods, the main functions of the shell are to guide the food-bearing water into the mantle cavity, limit the contamination of these nutrient-rich currents by expelled, waste-bearing water, and prevent sediment from entering the shell through the open valves. In most brachiopods the incoming feeding currents and discharged waste water are drawn in and expelled through separate parts of the shell. Specific flow patterns occur in brachiopods depending on the shell shape and orientation of the lophophore. The development of a central fold and sulcus in the brachiopod shell may have helped separate the incoming currents from the discharged waters, reducing the risk of refiltration of the exhalent waters.

Another modification seen in brachiopod shells is the development of a crenulated, or zig-zag, commissure. With the valves open this has the effect of increasing the length of the gape, and therefore the surface area of the mantle involved with respiration, without over opening the shell and allowing larger particles of unwanted suspended sediment to enter the mantle cavity.

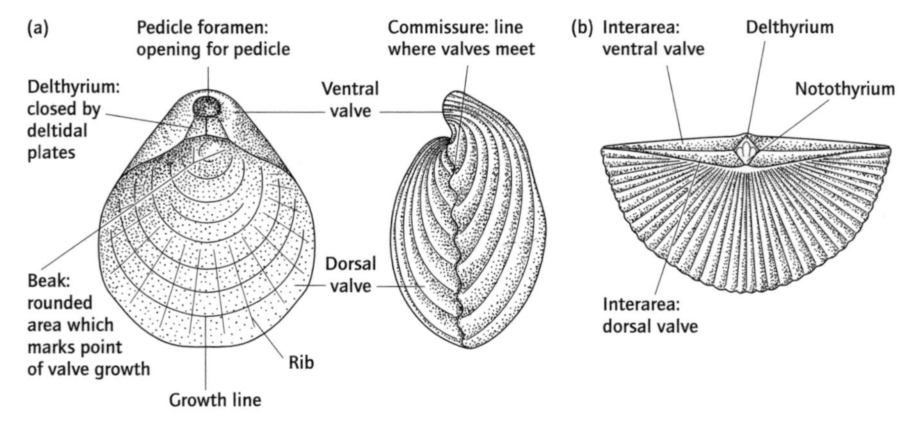

Fig. 6.2 Brachiopod external morphology: (a) astrophic, and (b) strophic.

Brachiopod ecology and palaeoecology

Brachiopods are exclusively benthic, marine animals. As filter-feeders they do not actively search for food and most brachiopods live on, or partially enclosed by, the substrate. They are dependent on currents to bring food and oxygen and carry away waste products. Most living brachiopods are attached to hard substrates, but fossil forms were much more diverse and exploited a range of benthic habitats, adapting their shell form and mechanism of attachment to suit the environment (Table 6.2).

Lingula, sometimes referred to as a "living fossil" because its morphology has not changed since the Ordovician, is able to burrow anterior first by using the pedicle as a prop as it enters the soft sediment. As it rotates and rocks its shell it forces its way through the sediment creating a U-shaped burrow so that it is in the correct orientation for feeding at the sediment–water interface, with the pedicle trailing below it. Other partial burrowers sat in the sediment, anchored by spines on the ventral valve or were stabilized by sediment covering the hinge.

Brachiopods with robust, strong pedicles (indicated by a large pedicle opening) required a hard substrate and were able to tilt the shell in order to maximize the current flow passing through the lophophore. The fold and sulcus present in some species may have separated water flow into incoming (nutrient rich) and outgoing (waste water) currents. Fossil brachiopods are also known to have encrusted and cemented themselves. The shell shape of encrusters reflected the nature of the attachment surface. Some brachiopods have a scar on the umbo of the pedicle valve showing the position where the brachiopod was attached, as a juvenile. Once the shell had grown enough to be stable the brachiopod became free living. Some Permian brachiopods had a morphology similar to a coral, with stilt-like spines attached to the apex of the ventral valve. These spines were either embedded in the sediment or cemented on to hard substrates, and stabilized the shell. The dorsal valve was reduced in such forms to a circular lid covering the ventral valve. These forms lived in densely packed, reef-like communities. Experimental models of these brachiopods show that by flapping the dorsal lid currents entered at the anterior region of the dorsal valve and waste was flushed out laterally. This theory is supported by the distribution of epifauna – commensal animals cemented on the shell. These were concentrated on the frontal area where currents were brought in. However, more recent work has suggested that this may be incorrect. A specimen has been found in apparent life position with the dorsal valve cemented to the substrate. Obviously this brachiopod would not have been able to flap this valve. Possibly the pumping action of the cilia drew water in and expelled waste products.

Unattached brachiopods, with a closed pedicle opening, must have lain on the sea floor, a rather unstable lifestyle. Some such brachiopods had large, saucer-shaped shells that helped distribute the weight evenly. Other adaptations included spines on the ventral surface that acted as anchors and spines along the commissure that helped to keep sediment out of the shell.

Table 6.2 Brachiopod lifestyle and morphology.

Mode of life	Shell form	Substrate	Example
Infaunal			
Burrowing	Smooth	Soft	*Lingula*
Semi-infaunal (partial burrowing)	Spines on ventral valve	Soft sediment overlying hard substrate	*Kochiproductus*
Epifaunal			
Attached by pedicle	Pedicle opening	Hard	*Magellania*
Encrusting	Closed pedicle opening, irregular ventral valve	Hard	*Crania*
Cementing	Closed pedicle opening, ventral spines, umbo scar	Hard	*Chonosteges*
Unattached	Closed pedicle, saucer shaped	Hard or soft	*Rafinesquina*

Community palaeoecology

Much work has been done on brachiopod communities. Palaeocommunities are fossil assemblages that represent the remains of a living community preserved where it lived. Fossil brachiopod communities are useful indicators of palaeoenvironment, in particular of the type of substrate and water depth. Many studies have used brachiopods to help understand the controls on faunal distribution and to reconstruct ancient environments. One of most famous of these is a study of the Lower Silurian in Wales, UK. Five communities, dominated by brachiopods, were identified and described in terms of species diversity and relative abundance. These communities formed concentric zones parallel with the palaeoshoreline (Fig. 6.3). Although distinct, they graded into each other across depths ranging from the intertidal zone into the deep basin. The nearest shore community was the least diverse and was dominated by *Lingula*, a linguliform brachiopod that lives on mud flats. The most distal community on the outer shelf was the most diverse. Each palaeocommunity is characterized by a particular brachiopod genus. It has been argued that their distribution was related primarily to water depth although other related factors such as pressure, salinity, and substrate quality may have influenced brachiopod distribution. An alternative suggestion is that temperature, which decreases with increasing depth, exerted the main control.

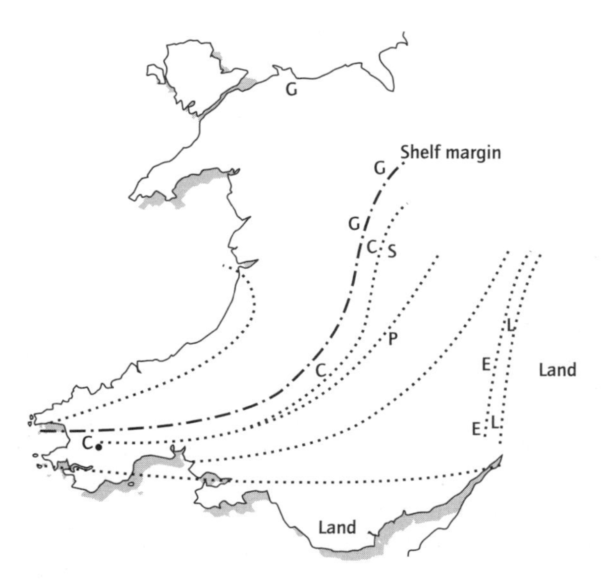

Fig. 6.3 Brachiopod palaeocommunities from Lower Silurian rocks: C, *Clorinda*; E, *Eocoelia*; G, graptolitic muds/shelf margin; L, *Lingula*; P, *Pentamerus*; S, *Costricklandia*.

Brachiopod evolution

Brachiopods originated in the early Cambrian; all three subphyla are known from this time and exist today. In the Cambrian, the Linguliformea and Craniiformea outnumbered the Rhynchonelliformea, although since this time the latter have dominated. In the early Ordovician the rhynchonelliforms underwent a massive radiation, possibly in response to the opening up of new habitats associated with continental breakup. They continued to flourish until the end of the Ordovician, when glaciation resulted in a significant decrease in diversity, although there were no major extinctions in the brachiopods. The radiation of rhynchonelliforms continued through the Silurian but the diversity of linguliforms and craniiforms was generally reduced.

A general brachiopod radiation occurred in the Devonian and they reached their maximum diversity in this period. However, by late Devonian times there was an overall decrease in diversity and abundance with significant orders of rhynchonelliforms becoming extinct. During the Carboniferous and Permian, rhynchonelliforms rediversified. This was a time when some of the most unusual and spectacular forms of brachiopods existed – brachiopods with a coral-like form and the semi-infaunal brachiopods with spiny ventral valves. However, brachiopods declined slowly through the Upper Palaeozoic and the end-Permian extinction, the largest mass extinction in Earth's history, removed the dominant Palaeozoic brachiopod groups.

The Mesozoic saw a general replacement of brachiopods by bivalves. The well-adapted siphon of the bivalves enabled them to exploit an infaunal mode of life, and to occupy environments inaccessible to brachiopods. Whether competition with the bivalves or the rise in brachiopod predators resulted in their decline, they never recovered their Palaeozoic status and only a few brachiopods made the transition to the modern fauna. Those that survive today tend to exploit harsh environments. *Lingula* lives on intertidal mud flats and most other living brachiopods are found in deep water.

Lingula

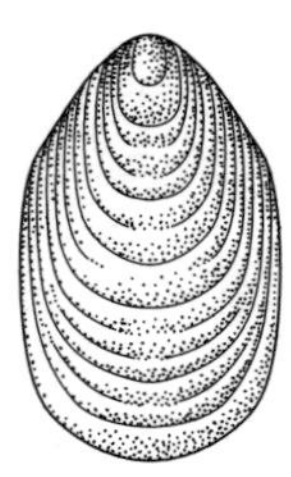

Linguliform brachiopod

Ordovician–Recent

A small (about 2 cm from the beak to the anterior edge), smooth, phosphatic brachiopod known as a "living fossil" as its morphology has not changed significantly since the Ordovician. Fully infaunal, it lives in burrows with its anterior edge close to the sediment–water interface. The pedicle anchors the brachiopod to the mud whilst the valves rotate and grind through the sediment. Modern *Lingula* mainly exploit marginal habitats, but fossil *Lingula* are known from shelf and basin environments.

Magellania

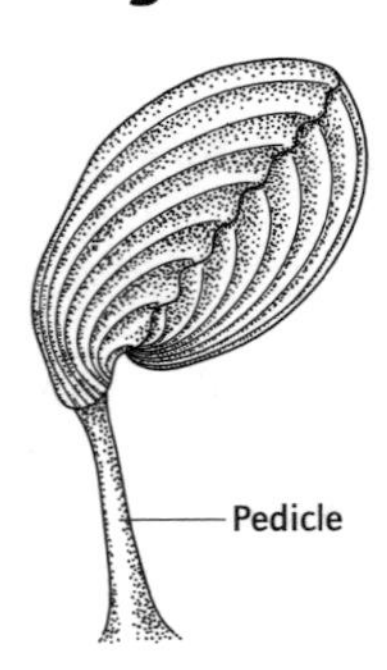

Rhynchonelliform brachiopod

Triassic–Recent

This Recent brachiopod lives in Australian and Antartic waters at a depth of between 12 and 600 m. It measures approximately 3 cm from the pedicle opening to the anterior edge. Using its stout pedicle it attaches to hard substrates. The pedicle acts as a stalk and raises the brachiopod from the attachment surface. The lophophore is supported by a distinctive calcareous loop attached to the hinge area of the dorsal valve. This structure is only rarely preserved in fossil brachiopods.

Gigantoproductus

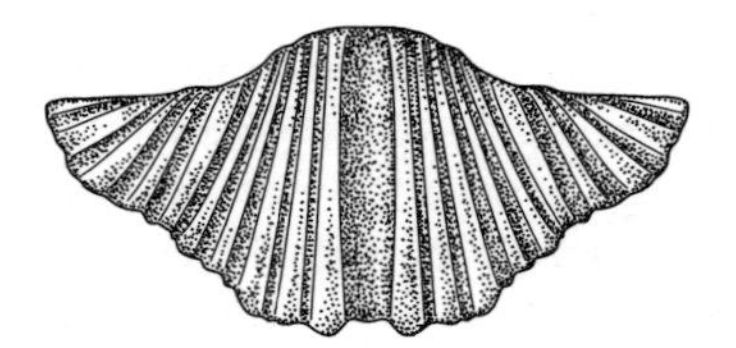

Rhynchonelliform brachiopod

Carboniferous

This uncommonly large brachiopod (up to 30 cm across the hinge) lived in very soft mud. The adult brachiopod lived semi-infaunally with the spiny ventral valve embedded in the sediment. The anterior margin curved upwards and grew very quickly to keep pace with sedimentation. This anterior margin is rarely preserved, diminishing the apparent size of the brachiopod.

Pentamerus

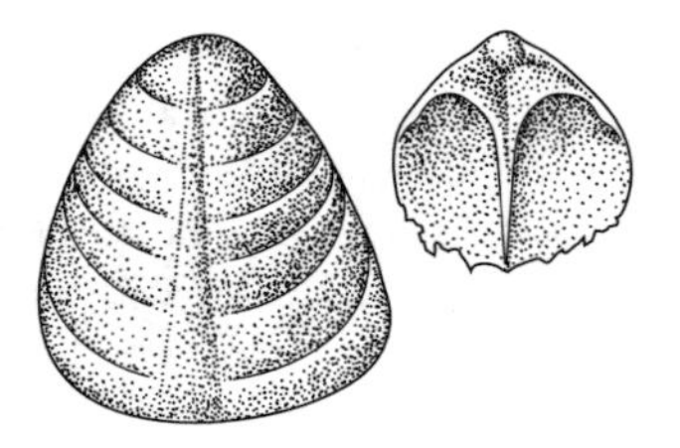

Rhynchonelliform brachiopod

Silurian

Pentamerus is one of the most common Silurian brachiopods. As both valves are strongly convex, it has a distinctive rounded shell with a hooked beak. It is typically 5 cm from the pedicle beak to the anterior edge. The wide delthyrium shows that it had a large pedicle. In the ventral valve muscles were attached to a characteristic muscle platform, called the spondylium, which raised them from the valve interior surface. This attachment is commonly preserved and the triangular shape of the anterior bisected by the vertical spondylium forms an arrow, which has given the local term "government rock" to Silurian sediments packed with *Pentamerus*.

Spirifer

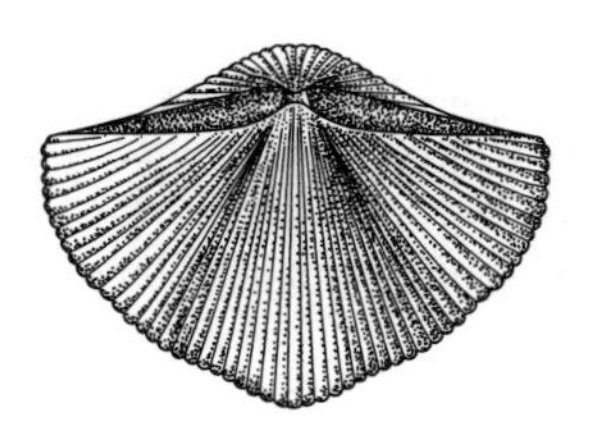

Rhynchonelliform brachiopod

Devonian–Permian

A triangular-shaped brachiopod with a long, straight hinge line (2–7 cm in length). The large delthyrium shows that the brachiopod lived attached to the substrate. The fold and sulcus enabled the brachiopod to separate the incoming nutrient-rich water from out going waste water. The triangular shape of the brachiopod is related to its internal morphology. The lophophore supports two symmetric spires with an axis parallel to the hinge line. These spiralia allowed the lophophore to filter water from the outside to the inside of the spire.

Prorichthofenia

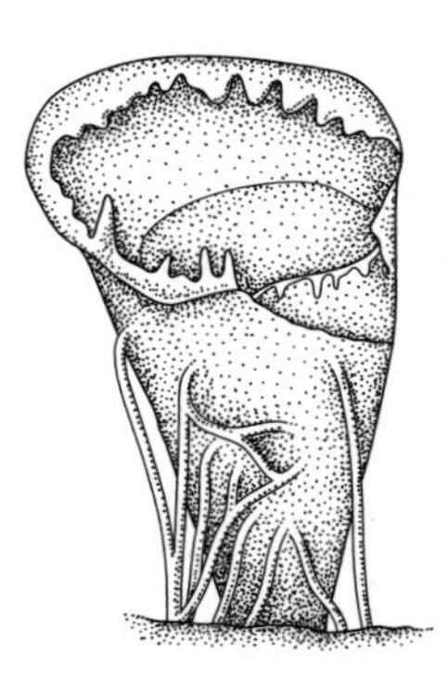

Rhynchonelliform brachiopod

Permian

One of the more unusual brachiopods with a coral-like form, approximately 6 cm high. Spines at the apex of the ventral valve rooted the brachiopod in the sediment and also fringe the anterior margin. The dorsal valve is almost circular (about 1.5 cm in diameter) and was recessed into the ventral cone. As with other cone-shaped brachiopods, *Prorichthofenia* formed reef-like communities.

Tetrarhynchia

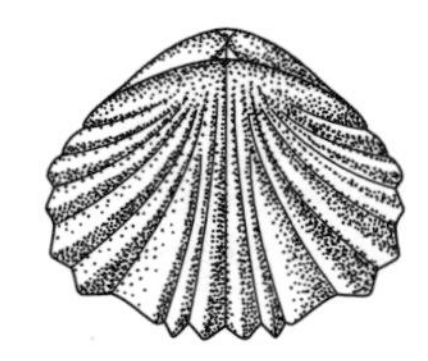

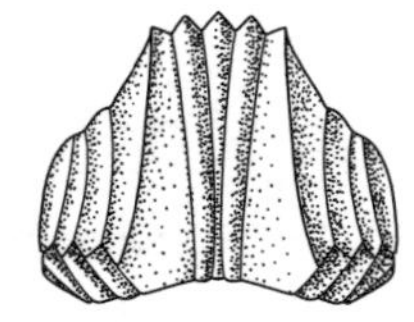

Rhynchonelliform brachiopod

Jurassic

This small (2 cm from ventral beak to anterior edge) brachiopod with a short curved hinge is strongly ribbed and has a characteristically deep fold in the dorsal valve and corresponding sulcus in the ventral valve. The zig-zag commissure enabled the brachiopod to increase the area open between the two valves, and hence increase nutrient intake, whilst restricting the entrance of larger particulate matter. The brachiopod was attached to the substrate by a small pedicle.

Colaptomena

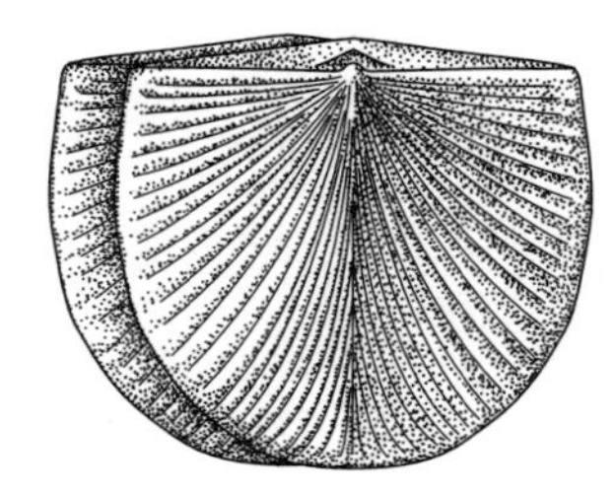

Rhynchonelliform brachiopod

Ordovician

A flattened, concavoconvex brachiopod with a strophic hinge line (approximately 2.5 cm in length). There is no pedicle opening and it lived unattached and on the sea floor with its curved surface facing upwards. When the valves were closed the internal volume was extremely low. The hinge was partially buried in the sediment to help stabilize the brachiopod.

Glossary

Adductor muscles – muscles that contract to close the valves.
Astrophic – curved hinge line.
Beak – point from which the valve starts to grow.
Brachidium – lophophore support.
Commissure – line along which the valves meet.
Delthyrium – triangular opening in the ventral valve.
Deltidial plates – plates that close the delthyrium.
Deltidium – single plate that closes the delthyrium.
Diductor muscles – muscles that contract to open the valves.
Dorsal (brachial) valve – valve to which the lophophore attaches.
Fold – elevated area in the anterior of the valve with a corresponding sulcus in the other valve.
Foramen – opening for the pedicle.
Interarea – area between the beak and the hinge line.
Lophophore – brachiopod feeding and respiratory mechanism.
Mantle – body tissues that secrete the shell.
Notothyrium – triangular opening in the dorsal valve.
Pedicle – fleshy stalk used to attach the brachiopod to the substrate.
Spiralia – coiled brachidia, typical of *Spirifer*.
Spondylium – muscle platform, typical of *Pentamerus*.
Strophic – straight hinge line.
Sulcus – depression in the anterior of the valve with a corresponding fold in the other valve.
Umbo – rounded area around the beak.
Ventral (pedicle) valve – valve to which the pedicle attaches.

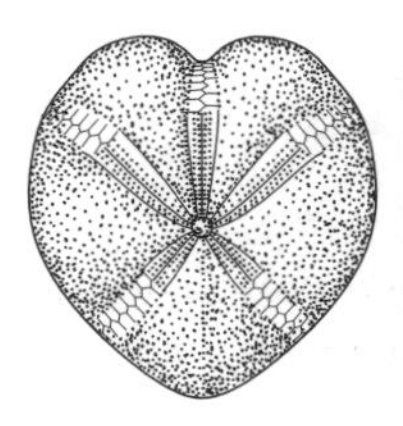

7 Echinoderms

- Echinoderms are diverse animals with a unique hydraulic system.
- Although their classification is problematic, echinoderms can be divided into six main classes.
- Most fossil crinoids lived attached to the substrate by a stalk. Living crinoids are usually stemless.
- Stellate echinoderms, starfish, and brittle stars, have a sparse fossil record but are the most common living forms.
- Echinoid morphology strongly reflects life habit.

Introduction

Echinoderms are diverse, commonly fossilized, marine animals. Living echinoderms include starfish, sea cucumbers, and sea urchins. Fossil forms are even more disparate. Echinoderms possess a number of distinctive features.

1 *Water vascular system*: This unique internal mechanism controls most echinoderm activity. Seawater is transported within the animal through a system of radial canals. Tube feet are the main organs for locomotion. They are also adapted for food collection, respiration, and can act as sensory tentacles.

2 *Endoskeleton*: Echinoderms have an internal skeleton (the test) formed of small porous plates (ossicles). Each ossicle is a single crystal of high magnesium calcite formed from a complex three-dimensional framework of rods (stereom). Spaces within this lattice are filled with soft tissue. Echinoderm ossicles are easy to identify in thin section as the presence of the soft tissue gives them a speckled appearance, and the single crystals of calcite forming individual ossicles have uniform optical properties. Ossicles are joined by living tissue to form the test, allowing the animal to grow without molting.

3 *Symmetry*: Many echinoderms have fivefold symmetry, although some groups have an imposed, secondary bilateral symmetry. This radial symmetry is usually associated with animals that have evolved from sessile groups that collect food from all sides. Why echinoderms evolved a five-rayed (pentaradiate) symmetry is not known.

The water vascular system and echinoderm lifestyle

Echinoderms are an extremely diverse group displaying a varied range of feeding strategies and lifestyles. Much of this diversification is attributable to the combination of radial symmetry and mobility. Radially symmetric animals are able to interface with the environment on all sides and are therefore usually suspension-feeders or passive predators. Echinoderms with their pentamerous radial symmetry show these modes of life. Their water vascular system also provides mobility, allowing echinoderms to live as infaunal deposit-feeders, epifaunal grazers, or active predators.

The water vascular system consists of fluid-filled canals with appendages, tube feet (podia) that project through porous plates (ambulacra) in the echinoderm endoskeleton. They are used for locomotion, food collection, respiration, and may perform sensory functions. In most echinoderms seawater enters the internal canals thorough a specialized porous plate, the madreporite. The fluid held within the water vascular system is a mixture of body fluids and seawater. The tube feet are extended by a hydraulic system that draws fluid from the main internal canals. It is not possible for all the tube feet to be extended or withdrawn at the same time. The correlation between the structure of the water vascular system and the echinoderm lifestyle is shown in Table 7.1.

Table 7.1 Relationship between the water vascular system and echinoderm mode of life.

Water vascular system	Lifestyle
Crinoids (sea lilies)	
The water vascular system in crinoids does not connect with the outside and the canals are entirely filled with body fluids. Canals extend from a central ring canal along each arm. In living crinoids the number of arms can be up to 200 and they may be branched. Internal canals run along each arm and the arm side branches (called pinnules). Tube feet project through the upper surface. They are highly mobile and perform feeding and sensory functions	Crinoids live attached to the substrate with their arms and pinnules outstretched, arranged in a fan perpendicular to the water flow. The tube feet are used in suspension-feeding. They intercept food particules and pass them into food grooves that lead to the mouth. This may be the original function of the water vascular system
Asteroids (starfish)	
In starfish seawater is drawn into the water vascular system through the madreporite. Each arm has a radial canal that bears tube feet, which project through the lower surface of the starfish. The tips of the tube feet are often flattened forming a sucker. The tube feet are primarily used for locomotion and holding prey during feeding	Most starfish are scavengers or opportunistic carnivores. Carnivorous starfish evert a portion of their stomach to consume their prey. The tube feet hold the starfish and the prey in position for the long periods necessary for digestion. Starfish living as suspension-feeders capture food with their tube feet
Ophiuroids (brittle stars)	
Brittle stars have a water vascular system similar to starfish but their tube feet are very different, they are more elongated, suckerless, and secrete a sticky mucus. In this group tube feet are functional in feeding, digging and also act as sensory organs	Brittle stars have a diversity of feeding strategies. Some brittle stars use their tube feet to extract organic material from the sediment. Many species are suspension-feeders. Some secrete a network of mucus thread to trap suspended food particles whilst others use the tube feet to intercept organic material. Predatory brittle stars use the podia to dig shallow, mucus-lined burrows from which they trap prey
Echinoids (sea urchins)	
Sea urchins have a similar system to starfish but the lateral canals extend around the sides of the body. It is as though the arms of a starfish have been tied together at the tips with the lower surface of the arms facing outwards. Sea urchins have specialized plates with holes through which the tube feet project. The tube feet serve many functions including locomotion, feeding, and attachment	Sea urchins exploit a wide range of lifestyles. Some scrape algae from the substrate, others live in burrows and use specialized tube feet to sort food material from the sediment and pass it to the mouth. Sand dollars use mucus-coated tube feet to extract food particles and some species use the tube feet to catch suspended organic material
Holothurians (sea cucumbers)	
The water vascular system of sea cucumbers is similar to other echinoderms but it is organized to accommodate an elongated body. It is as though a sea urchin has been stretched vertically and laid on its side with the mouth facing laterally. Radial canals extend along the upper and lower surface of the body, although in some species the tube feet on the upper surface are lost	Most sea cucumbers are suspension- or deposit-feeders. Suspension-feeders use branched tentacles to trap food particles. Deposit-feeders crawl across the substrate, using the tube feet to ingest organic-rich sediments. A few deep water forms have elongated tube feet for walking. Some cryptic species use the tube feet to anchor them in concealed positions

Classification

The classification of echinoderms is problematic. Essentially, they may be divided into two distinct subphyla: the pelmatozoans (fixed forms) and the eleutherozoans (mobile echinoderms) (Table 7.2). The most important pelmatozoans are crinoids. The main groups of eleutherozoans are starfish, brittle stars, sea urchins, and sea cucumbers.

Table 7.2 Characteristics of the main echinoderm classes.

Pelmatozoans	
Class Blastoidea	Echinoderms usually attached to the substrate by a stem. The calyx is bottle-shaped and contains specialized respiratory structures. Food-gathering appendages (the brachioles) are small, unbranched, and attached to plates on the outside of the calyx. Mid-Ordovician–late Permian
Class Crinoidea (sea lilies)	Generally stalked echinoderms with a bowl-shaped calyx covered by a flat tegmen. Food is collected by arms that are attached to plates in the upper part of the calyx. Arms may be branched and can bear lateral pinnules. Ordovician–Recent
Eleutherozoans	
Class Asteroidea (starfish)	Stellate echinoderms usually with five arms. Rows of large tube feet extend from an indistinct central disc along the lower surface arms. The mouth is situated on the lower surface. Asteroids are predators or scavengers. Early Ordovician–Recent
Class Ophiuroidea (brittle stars)	Star-shaped echinoderms with five slender arms extending from a distinct, large, central disc. Tube feet and mouth are situated on the lower surface. Early Ordovician–Recent
Class Echinoidea (sea urchins)	Echinoderms with the body encased in globular or flattened test, formed from many, fused calcite plates. Tube feet emerge through five rows of porous (ambulacral) plates. The mouth is usually situated on the underside of the test. Ordovician–Recent
Class Holothuroidea (sea cucumbers)	Bilaterally symmetric echinoderms with a muscular, elongated body. The endoskeleton is very reduced; only a few small ossicles are embedded in the body wall. Early Ordovician–Recent

Evolutionary history

In spite of their excellent fossil record, the evolutionary history of echinoderms is unclear. Some studies of the comparative morphology of echinoderm larvae have been used to establish the evolutionary relationships within the phylum. However, similarities between the larvae of ophiuroids and echinoids, for example, do not necessarily suggest that these two classes have a particularly close relationship but demonstrates that in both classes the larvae have acquired the same adaptations in response to selection for a planktonic mode of life. Therefore larval similarity is not necessarily indicative of phylogenetic relationships in adult echinoderms.

Echinoderms are known from the Lower Cambrian. The first true echinoderms may have been the spirally plated helioplacoids. These animals lacked the pentamerous radial symmetry characteristic of all other echinoderm groups. However, the endoskeleton was formed from ossicles with the unique stereom structure shared by later echinoderms and therefore helioplacoids may be considered the ancestral group of all echinoderms. A major dichotomy arises from this group. Two lineages result that eventually separate the crinoids from the other classes of echinoderms, the asteroids, ophiuroids, echinoids, and holothurians. Crinoids are fixed suspension-feeders whilst in the other echinoderm groups the tube feet are used primarily for locomotion. Maximum echinoderm diversity occured in the mid Palaeozoic. Crinoids appeared in the early Ordovician and were the dominant echinoderm species in the Palaeozoic. Asteroids and ophiuroids originated at the same time. There have been no major morphological innovations in these classes since the Palaeozoic. Echinoids underwent a major diversification in the Mesozoic. Holothurians are thought to be closely related to echinoids but their fossil record is sparse and their relationships are consequently difficult to determine.

Crinoid morphology

Crinoids originated in the early Ordovician and persist to the present day. Their maximum abundance was in the Palaeozoic. Most fossil crinoids were attached to the substrate by a stalk and occupied shallow water environments (Fig. 7.1). Modern species are more widely distributed living in habitats ranging from tropical reefs to cold, deep waters at polar latitudes. Reef-dwelling crinoids are stemless, are able to crawl and swim, and live in dense aggregations. Less abundant, deeper water forms, have a stem and resemble fossil crinoids.

All living crinoids are passive suspension-feeders. The arms and pinnules are arranged into a filtration fan with the mouth facing downcurrent (Fig. 7.2). Fine food particles are trapped by the tube feet, which form a fine net across the fan, and are passed to the mouth. Analysis of arm branching patterns of coexisting living crinoid species shows that different branching styles gather different-sized particles. Early crinoids with simple, unpinnnulated arms were probably unable to form filtration fans. Most Palaeozoic crinoids were less flexible than modern forms and were, therefore, probably unable to alter the orientation of their fan.

Stems raise the filtration fan away from slower moving water close to the substrate upwards into a zone where water moves faster. Some modern stemless crinoids feed in the faster currents by attaching to coral heads or prominent rocks.

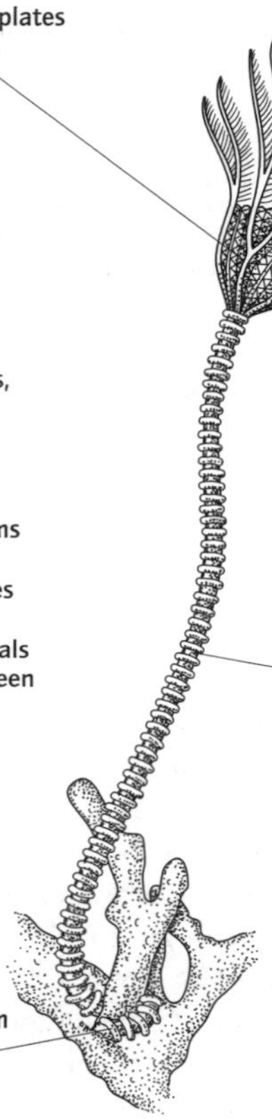

Calyx: connects the stem and the arms and contains the vital organs. The calyx is formed of a series of plates called radials, basals, and infrabasals

Pinnules: unbranched extensions that increase the surface area and efficiency of the filter fan

Monocyclic crinoids: have one series of plates between the stem and the radials, the basals

Dicyclic crinoids: have an additional series, the infrabasals, between the basals and the stem

Arms: crinoid arms radiate from the calyx. Five arms are attached to the radial plates of the calyx. Usually arms branch at least once, at the base. Each arm consists of a series of brachial plates

In some crinoids the lower parts of the arms are incorporated into the calyx. These plates are known as fixed brachials. Interbrachials fill in the space between the brachials

Stem: formed from a series of ossicles called columnals. Columnals are joined by elastic ligaments and the stem is flexible. On death these ligaments quickly decay and the stem disintegrates into individual ossicles that are a common component of limestones. Some columnals are smooth but others interlock and are adapted to resist stem torsion

Holdfasts: some crinoids were cemented to hard substrate, others were anchored by root-like structures. In this example the distal part of the stem coils around a fixed bryozoan

Fig. 7.1 Ordovician crinoid *Pycnocrinus*.

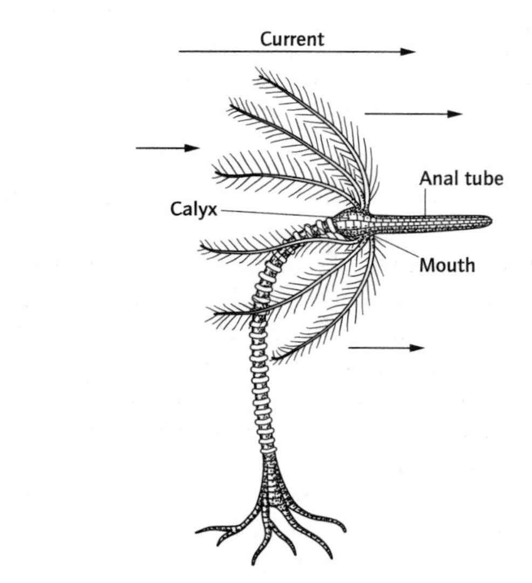

Fig. 7.2 Reconstruction of crinoid feeding.

Crinoid evolution

The first true crinoids appeared in the early Ordovician. Rapid diversification followed and all the main Palaeozoic crinoid groups were established by the middle Ordovician. Most Palaeozoic crinoids were attached to the substrate by stems. Extinction at the end of the period resulted in a depleted crinoid fauna in early Silurian times. Subsequently crinoids reradiated reaching their peak diversity in the early Carboniferous.

By the beginning of the Mesozoic, crinoid abundance and diversity had catastrophically declined. Most Palaeozoic groups became extinct and only a single group survived into the Mesozoic. During the Triassic, crinoids diversified and new groups were established. Stemless forms first appeared in late Triassic times, increasing in dominance in the Jurassic. Some groups were affected by the end-Triassic extinction event but since this time crinoids have continued to the present day without any major changes in diversity or abundance.

Asteroids

Asteroids, or starfish, are very distinctive echinoderms. Most have five arms and exhibit the classic echinoderm fivefold symmetry. The mouth and tube feet are positioned on the lower surface. Tube feet are arranged in rows along the arms and the mouth is central (Fig. 7.3). Starfish are voracious scavengers and predators, preying particularly on bivalves. Some species are able to evert their stomach. Specialized tube feet are used to prize bivalves apart. The stomach then protrudes into the shell digesting the soft tissues inside the shell.

The asteroid skeleton (test) is extremely flexible. It is formed from small calcite plates that are not tightly sutured together. Therefore, the test disarticulates rapidly after death and identifiable fossil starfish are very rare.

Starfish have been important predators for most of the Phanerozoic. Interlocking bivalve shells may have developed in response to the spread of predatory starfish in late Ordovician and early Silurian times. Starfish predation may also be responsible for the failure of brachiopods to reradiate after the end-Permian mass extinction.

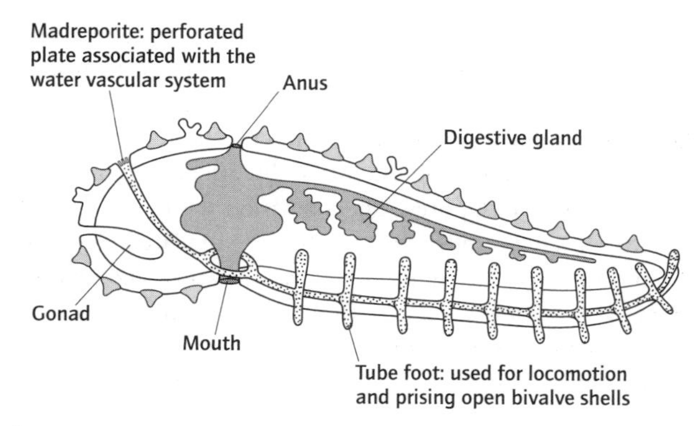

Fig. 7.3 Vertical section of a starfish.

Ophiuroids

Ophiuroids, or brittle stars, are star-shaped echinoderms with five slender arms radiating from a distinct, circular, central disc (Fig. 7.4). The arms are extremely flexible and are formed of specialized vertebrae-like plates. The mouth is on the lower surface at the center of the disc.

Ophiuroids gather food with their arms. They are very mobile and are able to coordinate their arm movements to allow for relatively rapid crawling and swimming. Some species live in shallow water, although most prefer deeper water environments (below 500 m) where there are fewer predators.

The fossil record of ophiuroids is poor. The skeleton fragments easily and complete specimens are extremely rare. Ophiuroids originated in early Ordovician times and their skeletal structure has remained generally unchanged since then. Modern ophiuroids are very diverse and form the largest living echinoderm class.

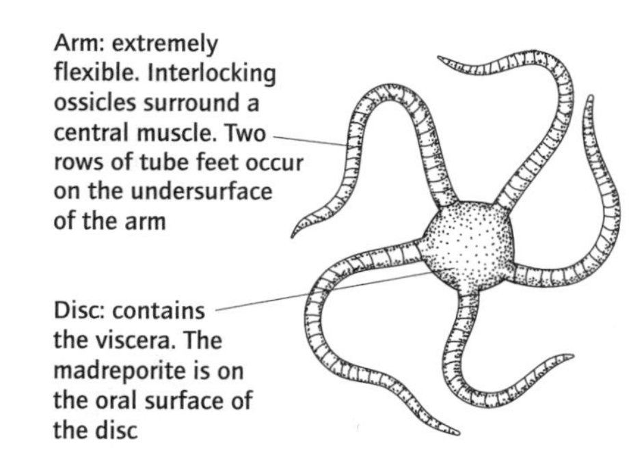

Fig. 7.4 Ophiuroid morphology.

Echinoid morphology

Echinoids, or sea urchins, have a robust, internal skeleton (the test) composed of numerous, fixed, calcite plates. Instead of arms the test has five narrow zones formed from perforated plates (the ambulacra) through which the tube feet emerge. These porous segments alternate with broader areas (the interambulacra) that lack pores. The anus is on the upper (aboral) surface and is surrounded by a double ring of plates (Fig. 7.5). The mouth is on the underside (oral surface). The external surface of the test is covered with spines and pedicellariae, tiny spines with pincers that remove settling organisms (Fig. 7.6).

Echinoids can be divided into two main groups: the regulars, rounded forms (e.g., sea urchins), and the irregulars, flattened and heart-shaped echinoids (e.g., sand dollars, heart urchins) (Fig. 7.5). Regular echinoids are always surface-dwellers and usually feed by scraping seaweed from rocks using a complex jaw apparatus known as "Aristotle's lantern". Articulated spines enable the animal to move slowly across the substrate aided by the tube feet. Irregular echinoids are often burrowers, and their spines are generally shorter and more densely spaced. The tube feet are highly modified. Some are used for digging and others are adapted for respiration, forming tubes that connect the animal with the sediment surface.

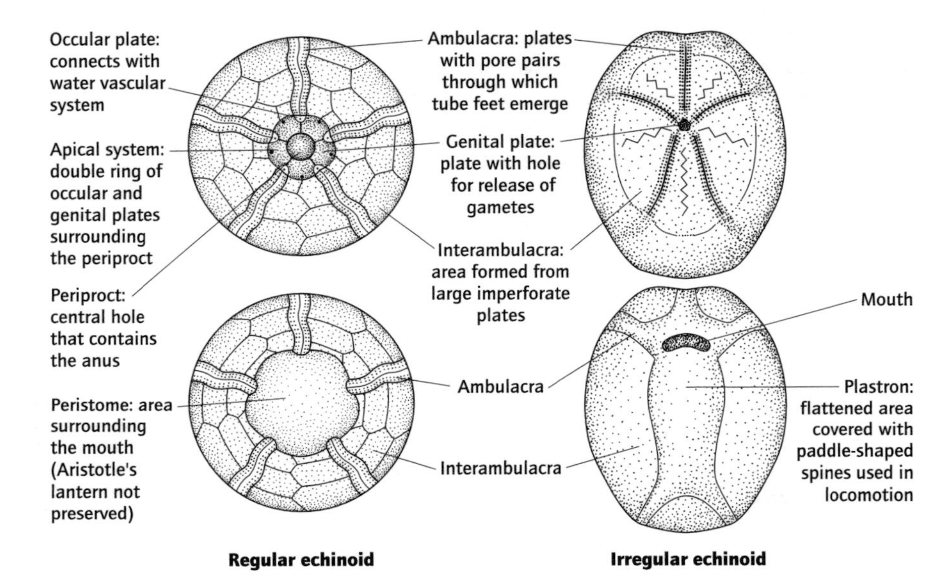

Fig. 7.5 Regular and irregular echinoid morphology.

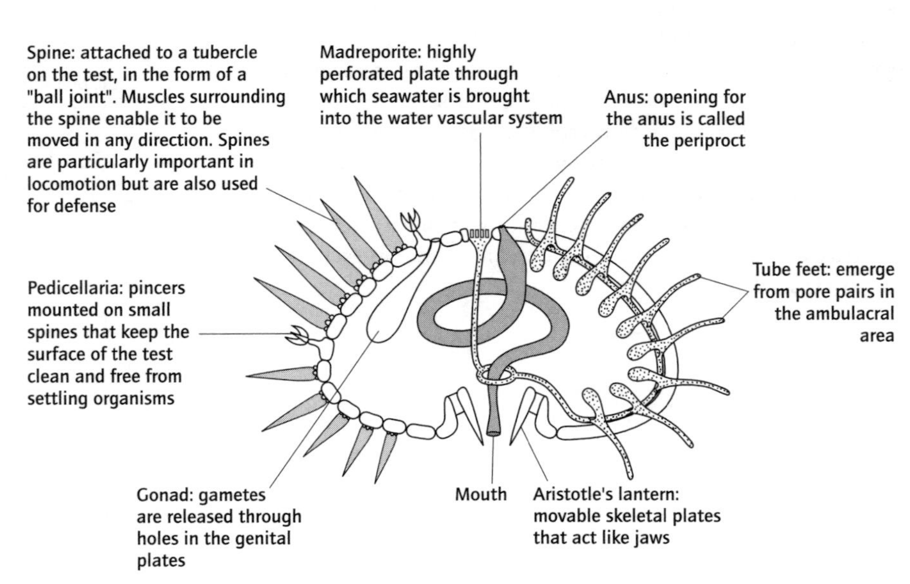

Fig. 7.6 Vertical section of an echinoid.

Echinoid ecology

Echinoids exploit three main life habits represented by three very distinctive morphologies (Table 7.3, Fig. 7.7).

Table 7.3 Echinoid life habits and morphology.

	Epifaunal	Shallow infaunal	Deep infaunal
Mode of life	Echinoids living on the substrate surface as scavangers or grazers in intertidal or shallow subtidal environments	Echinoids that are able to burrow rapidly in high-energy, shifting, sands	Echinoids that construct structured, semipermanent burrows in low energy environments
Morphology	Regular echinoids with a rounded test and radial symmetry. The anus is on the upper surface and the mouth directly opposite on the underside of the test	Irregular echinoids with very flattened, bilaterally symmetric tests. Ambulacra are petal shaped. The anus and mouth are on the lower surface	Irregular echinoids with heart-shaped, bilaterally symmetric tests. Ambulacra are petal shaped. The anus is on the posterior margin of the test

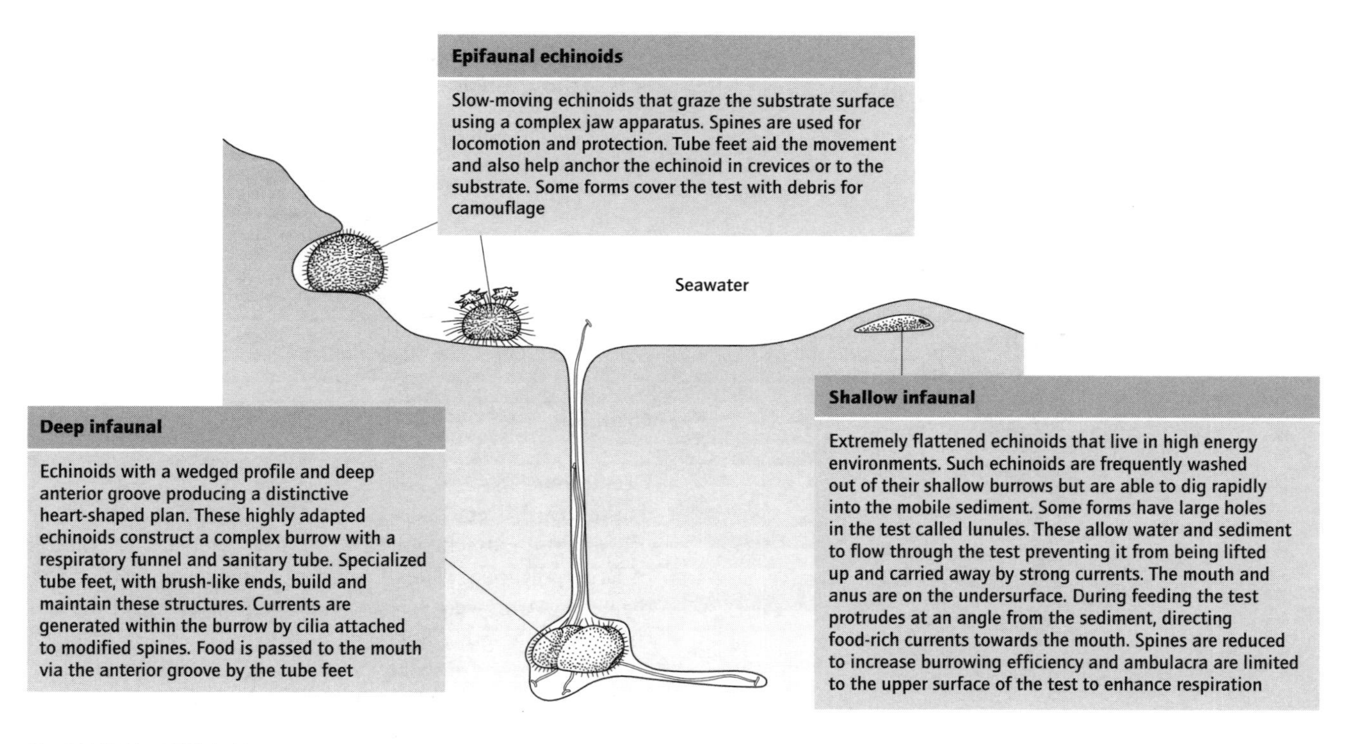

Fig. 7.7 Echinoid life habits.

Echinoid evolution

Although regular echinoids are known from the Ordovician they are generally uncommon in the Palaeozoic. Lower Palaeozoic forms tended to be small. Size generally increased through the era. Echinoids declined significantly in the late Carboniferous and only a few groups survived the end-Permian mass extinction event.

Echinoid abundance increased in early Mesozoic times. The group underwent a major radiation in the early Jurassic. Irregulars first appeared during this period, and by the Cretaceous echinoids were exploiting a wide range of infaunal habitats. Flattened sand dollars first appeared in the Palaoecene and these highly modified echinoids quickly became widely distributed.

Amphoracrinus

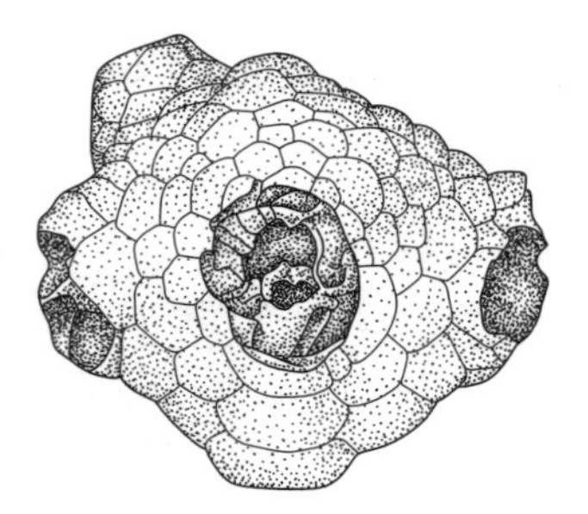

Crinoid

Lower Carboniferous

This is the calyx of the Carboniferous crinoid *Amphoracrinus* (height approximately 4 cm). Three basal, five radial, and five brachial plates form the lower part of the calyx. A plated, domed tegmen covers the calyx and there is a pronounced anal tube. Although only rarely preserved, the arms are known to branch many times. The whole calyx structure is rigid.

Pentacrinites

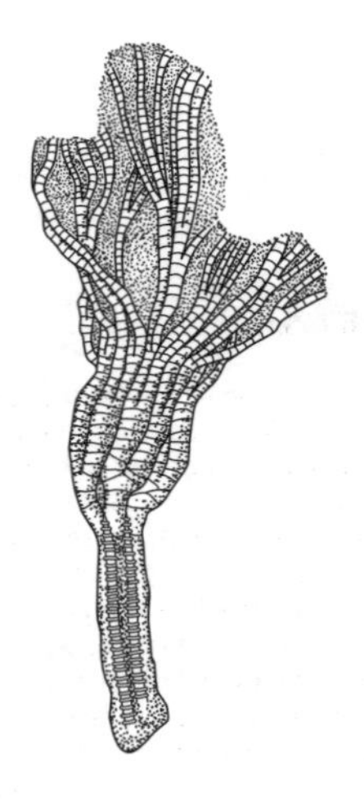

Crinoid

Triassic–Recent

This crinoid has a small calyx and very long arms. The length of the figured specimen is approximately 13 cm. Exceptionally preserved specimens of *Pentacrinites* are found in the Lower Jurassic marls of Dorset, UK. These crinoids are commonly associated with fossil driftwood. It has been proposed that this species of *Pentacrinites* was pseudoplanktonic, living attached to floating driftwood. Eventually the wood became waterlogged and sank into anoxic sediments that preserved the crinoids.

Apiocrinites

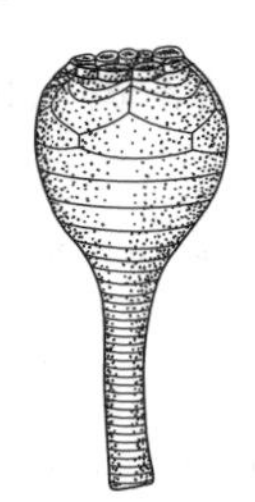

Crinoid

Jurassic–Recent

Apiocrinites has a distinctive barrel-shaped calyx (approximately 4 cm in height). The stem is formed from robust columnals and can be up to 15 cm in length. It swells at the base forming a holdfast that cemented the crinoid to the substrate. Such crinoids are associated with marine hardgrounds and have been found with encrusting organisms. Often only the robust holdfasts of these crinoids are preserved.

Marsupites

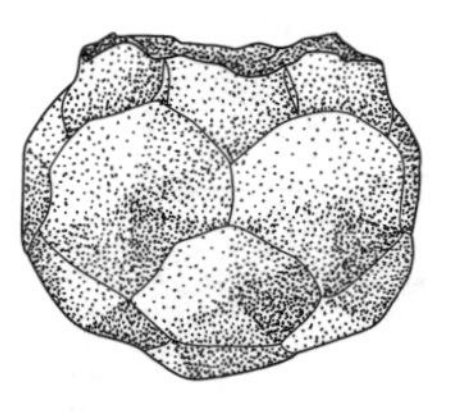

Crinoid

Cretaceous

This unusual Cretaceous crinoid does not have a stem or any other form of attachment. The calyx is globular and consists of large polygonal plates (mature specimens can be 6 cm in diameter). Arms were probably very long (up to 1 m) although their precise length is unknown.

Marsupites was probably benthic, living with the calyx embedded in the soft chalky sediment of the late Cretaceous sea beds.

Archaeocidaris

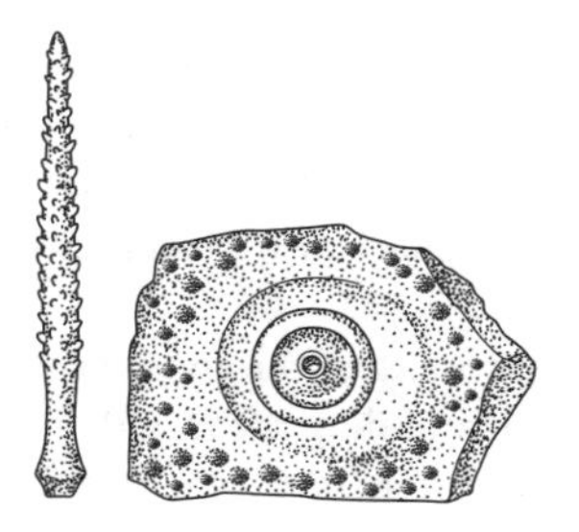

Echinoid

Lower Carboniferous

Archaeocidaris belongs to the subclass of echinoids Cidaroidea. Cidaroids are the only Palaeozoic echinoids to survive to the present day. The test was broadly hemispherical with narrow sinuous ambulacra. The interambulcral plates (shown above) had a prominent, central tubercle. A long, robust spine (approximately 6 cm in length) was attached to the tubercle. Living cidaroids use the spines for walking.

Clypeus

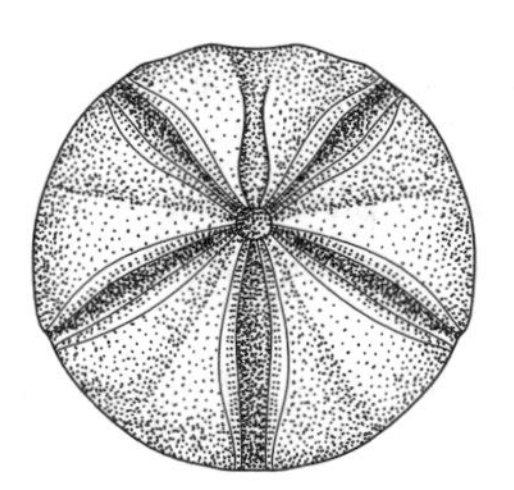

Echinoid

Middle–Upper Jurassic

This irregular echinoid has a flattened, bilaterally symmetric, discoidal test with a rounded outline (diameter approximately 6 cm). Ambulacral areas are expanded into the full petaloid condition with slit-like pores. The anus is situated within a pronounced groove on the aboral surface. The mouth is at the center of the oral surface.

Clypeus was an infaunal echinoid. The ribbon-like tube feet were adapted for respiration.

Hemicidaris

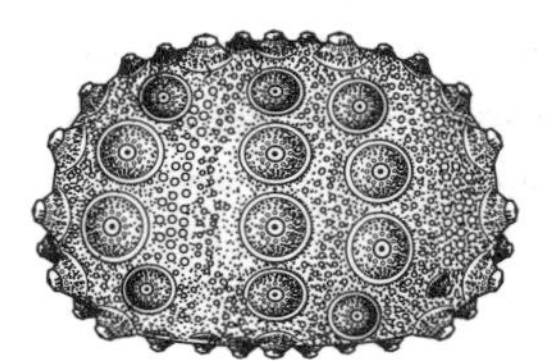

Echinoid

Middle Jurassic–Upper Cretaceous

This regular echinoid has a radially symmetric, hemispherical test (approximately 3 cm in diameter). The ambulacra are narrow and the plates are ornamented with small tubercles. The interambulacral plates have a large, central tubercle surrounded by smaller, less prominent tubercles. A long, solid, primary spine was articulated on the central prominence and shorter spines attached to the minor tubercles. Spines were used for protection and locomotion.

Such regular echinoids were epifaunal and lived in shallow marine and intertidal environments.

Micraster

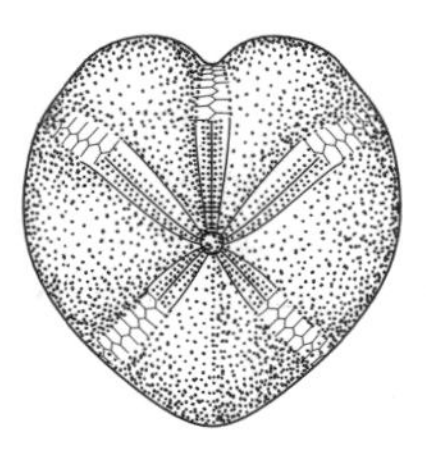

Echinoid

Upper Cretaceous

Micraster is an infaunal echinoid with a distinctive heart-shaped test (approximately 5 cm from posterior groove to anterior margin). Ambulacra are narrow and subpetaloid. The anterior ambulacrum is situated in a deep groove that leads towards the mouth. The anus is positioned on the anterior margin. Below the anus there is a small fasciole, a current-generating area that sweeps waste material into a sanitary tube. The mouth is on the lower surface, positioned towards the anterior margin. It is partially covered by a pronounced lip or labrum. Behind the mouth is a flattened area, the plastron.

Evolutionary changes in *Micraster* are well documented and changes in morphology can be related to the depth of burrowing.

Glossary

Ambulacra – plates with pore pairs through which the tube feet emerge.
Anal tube – prominent plated cone-like structure in crinoids that projects from the oral surface. The tube contains the anal opening.
Apical system – double ring of plates surrounding the periproct in echinoids.
Aristole's lantern – complex jaw-like mechanism usually found in regular, and some irregular, echinoids.
Basal plates – lower ring of plates forming the crinoid calyx.
Brachials – ossicles forming the arms in crinoids.
Calyx – plated cup-like structure in crinoids and blastoids containing the viscera. Also known as the theca.
Columnals – ossicles forming the stem in crinoids.
Dicyclic – crinoid calyces with an extra ring of plates, the infrabasals, between the basals and the stem.
Fasciole – areas of the echinoid test that generate currents. Common in infaunal echinoids.
Genital plates – one of two types of plates forming the apical system. Plates have a hole through which gametes are released.
Interambulacra – ossicles without perforations.
Irregular echinoids – bilaterally symmetric echinoids, usually infaunal.
Madreporite – specialized genital plate through which water is drawn into the water vascular system.
Monocyclic – crinoid calyces with only one ring of plates between the stem and the radials, the basals.
Occular plate – one of two types of plate forming the apical system. Plates are part of the water vascular system.
Ossicle – plate forming part of the echinoderm endoskeleton.
Pedicellariae – tiny spines with pincers that remove settling organisms in echinoids.
Periproct – area surrounding the anus in echinoids.
Peristome – area surrounding the mouth in echinoids.
Pinnule – simple side branches of the crinoid arms.
Plastron – flattened area behind the mouth in irregular echinoids with specialized paddle-shaped spines.
Radial plates – upper ring of plates forming the calyx with an articulation site for the arms.
Regular echinoids – echinoids with pentaradiate symmetry.
Stereom –microscopic lattice of rods permeated by tissue that forms the echinoderm plates.
Tegmen – cover for the oral surface in crinoids, sometimes soft, sometimes developed as a heavily plated "roof".
Test – echinoderm endoskeleton.
Theca – plated cup-like structure in crinoids and blastoids containing the viscera. Also known as the calyx.
Tube feet – lateral extensions of the radial water vessels used in locomotion, respiration, and food gathering.
Viscera – echinoderm main body organs.
Water vascular system – complex hydraulic system unique to echinoderms, principally used in locomotion and feeding.

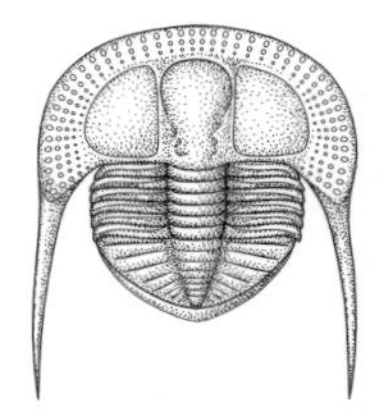

8 Trilobites

- Trilobites are arthropods, a phylum that includes insects and crustaceans.
- Between the Cambrian and the Permian they were amongst the most important elements of marine communities.
- Their body plan, with flexible segments and many limbs, allowed them to occupy a wide variety of ecological niches.
- Trilobites went extinct in the end-Permian mass extinction event, but had been in decline throughout the Upper Palaeozoic.

Introduction

Trilobites belong in the phylum Arthropoda, the most diverse phylum on the planet today. Arthropods include all insects, millepedes, centipedes, scorpions, and crustaceans, such as crabs and lobsters, in addition to a range of more obscure forms. If estimates of 10 million living species of insect are correct, then this phylum outnumbers vertebrates, for example, by a hundred times.

Arthropods all have a segmented body and many, jointed limbs (this is what gives the group its name). They have a well-developed head and sensory system, and probably evolved from an ancestral worm during the late Precambrian. The skeleton is external and is shed and replaced as the animal grows. This process, known as molting or ecdysis, is a major controlling factor in the success of the group. On the positive side, it allows a single individual to assume multiple body plans through life; for example, it allows caterpillars to turn into butterflies. On the negative side, it uses up costly resources and makes the animal helpless and prone to predation each time it occurs.

Trilobites radiated early in arthropod history, and were most abundant during the early Palaeozoic. However, recent work suggests that they were highly adapted and not a primitive root stock for other arthropods.

Although trilobites are well known, there are frustrating gaps in our knowledge. The main reason for this is that the exoskeleton on the upper body was mineralized, by the addition of calcite to an organic template, and so preserves well, but the lower body exoskeleton was not. Limbs and appendages were rarely mineralized, and then only very lightly. As a result, few trilobites have been preserved with legs, gills, or antennae, and the hypostome, a mineralized plate under the mouth, usually fell away as the animal decayed. The range of variation in the morphology of the underside of trilobites is largely unknown and most reconstructions rely on one of a handful of exceptional specimens.

Trilobites were mainly bottom-dwellers, and as such, tended to develop marked provincialism in their faunas. This makes them useful palaeogeographic indicators, especially in the Cambrian and early Ordovician. For rocks of this time period, trilobites are also useful for biostratigraphy, although in younger rocks the typical duration of a species is too long to be of much value.

Trilobite morphology

Trilobites were divided across the body into a head (or cephalon), thorax, and tail (or pygidium), and along the body into three lobes – a central axial lobe covering the main body cavity, and two pleural lobes covering the legs and gills on either side (Fig. 8.1). The cuticle from which the trilobite exoskeleton was constructed was layered, with a thin outer layer and a thicker inner one. Both were made from calcite, arranged in an organic matrix which has not yet been characterized.

The head, or cephalon, was generally large relative to the rest of the body, and was usually divided into a series of sections separated by sutures that facilitated molting. These sutures form a small number of distinctive patterns across the cephalon. The section of the head outside the facial sutures is known as the free cheek. The part inside, adjacent to the glabella, is called the fixed cheek. The eyes of the trilobite could attach to either the free or fixed parts of the cephalon.

The eyes of trilobites were typically large and compound, made from multiple lenses like the eyes of flies. Eyes in arthropods have evolved several times, and those in tribolites are unique to the group. They had very clear vision, often using the physical characteristics of calcite to enhance light capture and focus. The visual range of trilobites was very variable, and appears strongly related to their mode of life. Free-swimming forms had forward-facing eyes, sometimes organized in a single, continuous band around the front of the head. Burrowing forms often had raised eyes close to the top of their cephalon.

Running down the middle of the head was the glabella, a raised region that protected the stomach. The size of the glabella correlated to the size of the stomach, and furrows that partly run across the glabella may be the external expression of internal ridges, to which ligaments supporting the stomach were attached. The mouth was situated underneath the cephalon, near the back of the head, and is associated with a plate of cuticle called the hypostome. This could be attached to the front of the cephalon, or free lying within the soft cuticle of the underside of the trilobite (Fig. 8.1c).

The thorax of a typical trilobite was made up of a series of nearly identical segments, usually between two and 20 in number. Below each of these was pair of legs and gills. In Ordovician and later trilobites these segments were usually jointed in such a way that the trilobite could roll up for defense (Fig. 8.1b). Sometimes structures on the head and tail locked together to make enrollment even more effective.

The trilobite tail, or pygidium, was usually small and made up of a series of fused segments that look similar to those on the thorax. It is likely that leg/gill pairs occurred under some segments of the tail in most trilobites, though in some species these segments may have lacked limbs.

Three unusual evolutionary directions were taken by trilobites, taking them away from this rather conservative body plan. One involved them becoming extremely spiny. A second involved secondary loss of eyes, and sometimes the addition of highly pitted fringes around the front of the head. The third involved a great reduction in size, and in the number of thoracic segments to only one or two (Fig. 8.1d). Each of these adaptations can be related to changes in lifestyle (see p. 56).

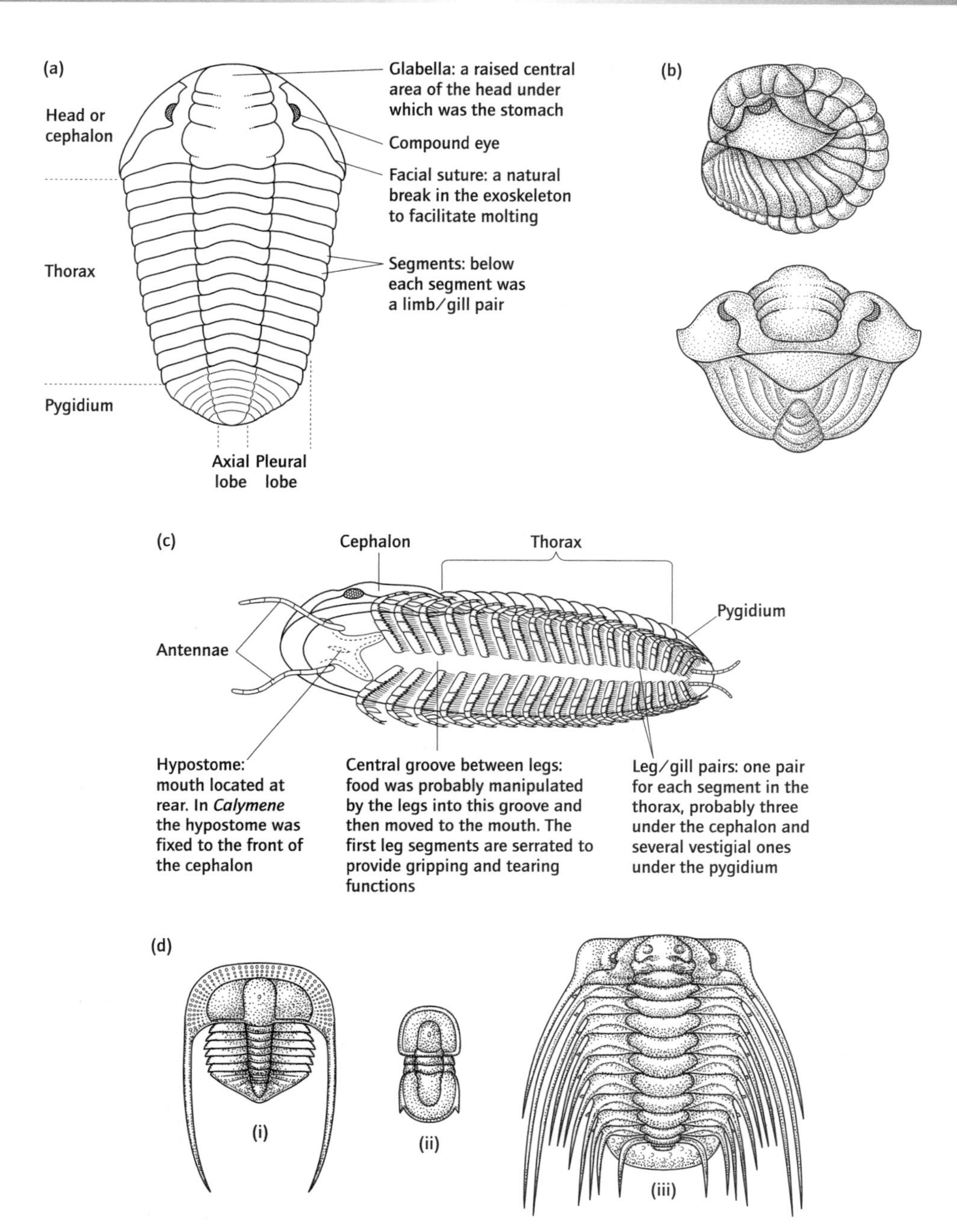

Fig. 8.1 The main elements of trilobite morphology. (a) *Calymene*, a Silurian predatory trilobite in dorsal view, showing the main, well calcified, elements of the carapace. (b) Two views of the same animal enrolled, showing the tight fit that was made between the front of the cephalon and the back of the pygidium. (c) *Calymene* as it might have looked from below, showing the lightly calcified or organic skeleton including the hypostome, legs, and gills. (d) The three main adaptive strategies of trilobites away from a highly conserved body plan: (i) *Trinucleus*, a blind trilobite with large frontal pitted region, which probably had a sensory function; (ii) *Agnostus*, a tiny trilobite with a much reduced thorax; and (iii) *Selenopeltis*, a representative of the extremely spiny adaptation of trilobites.

Trilobite mode of life

A great deal of work has been done on determining the mode of life of trilobites. Evidence for these deductions comes from the distribution of particular species of trilobites, modifications to their shape, and physical experiments conducted on model trilobites. The result is that many trilobites can be confidently located in terms of the depth of water in which they lived and the ecological role that they occupied. Several elements of trilobite morphology are useful in helping to infer life habit, especially the hypostome.

The earliest lifestyle adopted by trilobites was probably a predatory one. These animals ate worms and other soft invertebrates and had a rigidly attached hypostome and spiny bases to their limbs (as in Fig. 8.1c). Deposit-feeders, extracting food from soft sediment, and filter-feeders, extracting food from the water, evolved from these carnivorous ancestors. In deposit-feeders the hypostome tends to be detached from the cephalon. Filter-feeding trilobites functioned by suspending sediment in water using their legs, and then drawing the water into their partly enrolled body and extracting food from it. These trilobites changed the shape of their cephalon, and sometimes of their whole body, to form a large chamber under the carapace within which sediment could be suspended. The hypostome is often set back into the cephalon, at an unusual angle.

All trilobites were mobile and most seem to have crawled on the sea bed. A small number of species became burrowers or active swimmers. Most extremely, some trilobites became pelagic, with a permanently swimming, active lifestyle above the sea bed. These forms had eyes with up to 360° vision and a streamlined body. The most bizarre lifestyle adopted by trilobites is seen in a group of genera exemplified by *Olenus*. These trilobites had flattened bodies and many thoracic segments. They often occur in black shales, indicative of low oxygen availability above the sea bed, and had a great abundance. Each thoracic segment would have had a gill, allowing maximum extraction of oxygen, and it may be that they farmed sulfate-reducing bacteria, as do modern animals living in similar conditions near black smoker vents. Most of these lifestyles are shown in Fig. 8.2.

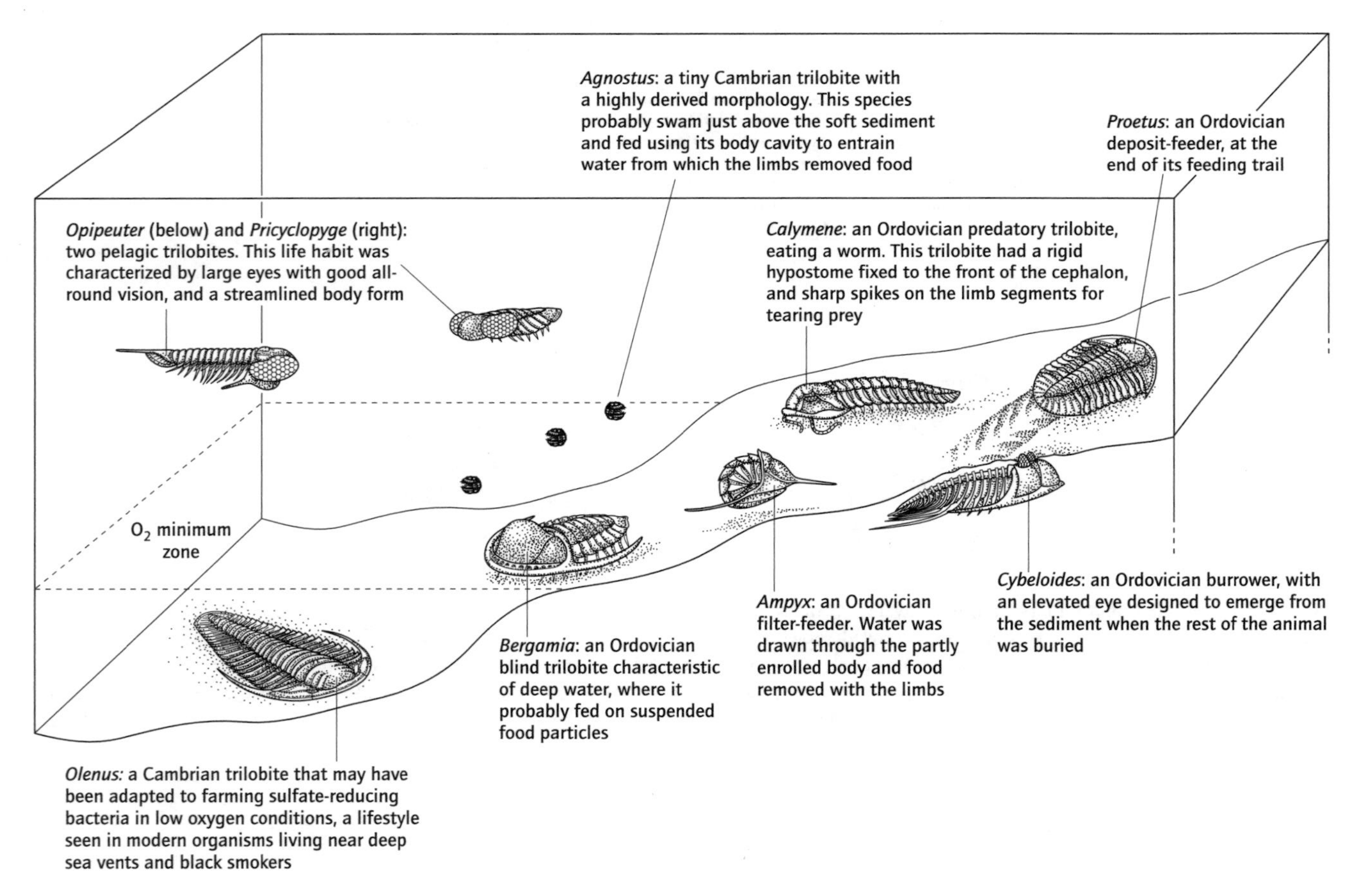

Fig. 8.2 A reconstruction of a range of late Cambrian to Silurian trilobites showing their probable life habits and positions. The reconstruction deepens from right to left, and there is an oxygen minimum zone above the sea bed in the extreme left of the image. All trilobites were marine.

Trilobite evolution

Trilobites were a major part of the marine benthos for over 250 million years. During all of that time their basic body plan remained the same, and the changes that did occur through evolution tended to be in details rather than in serious shifts of shape. This lack of innovation in a successful group is known as evolutionary conservatism. However, despite this, trilobites inhabited a wide range of niches and explored a wide range of marine environments from their evolutionary origins in the Cambrian.

Cambrian trilobites showed high diversity and included tiny, blind forms – the agnostids – and a range of more "familiar" body shapes. Some Cambrian trilobites show a secondary loss of eyes. These blind forms probably lived in deep water. A common factor amongst Cambrian trilobites is that they seem to lack any adaptations for defense against predators. A major extinction in late Cambrian times may have been linked to the appearance of common, large predators, especially molluscs. Species that radiated after this extinction show a diverse array of defense strategies, including the ability to enroll, or burrow, or extreme spinosity.

This early Ordovician radiation saw trilobites diversifying into the wide range of niches generated by the Palaeozoic sea bed community (see Chapter 16). This is the period of greatest diversity of tribolite body forms, although more species have been recognized from Cambrian rocks. In common with many Palaeozoic organisms, trilobites were badly affected by the sudden glaciation which caused the end-Ordovician mass extinction.

Recovery from the late Ordovician mass extinction was limited, and trilobite variety and diversity remained subdued for the rest of their evolutionary history. During the late Devonian diversity fell to two families, though there was some recovery in the Carboniferous. The last surviving trilobites became extinct in the end-Permian mass extinction (Fig. 8.3).

The high diversity of trilobites in Cambrian and Ordovician rocks makes them biostratigraphically useful in rocks from these periods. This is especially true for the middle and late Cambrian, as Ordovician forms tend to be limited to particular facies and geographic regions. Later trilobites tend not to be used for biostratigraphy because of their relatively long ranges. However, some local schemes have been erected based on their occurrence.

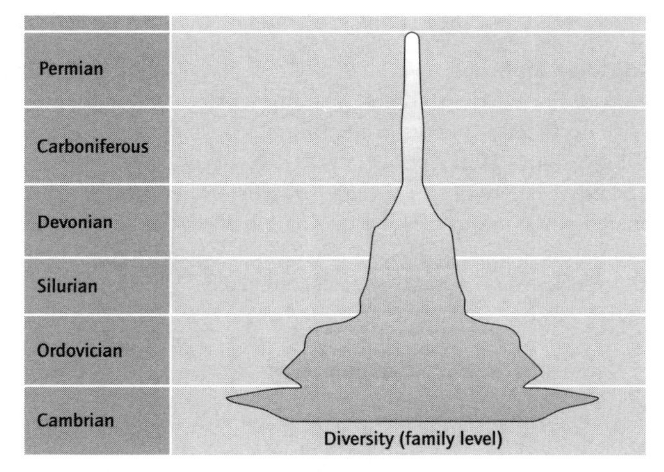

Fig. 8.3 Trilobite diversity through time. Note the high diversity in the late Cambrian and early Ordovician.

Paradoxides

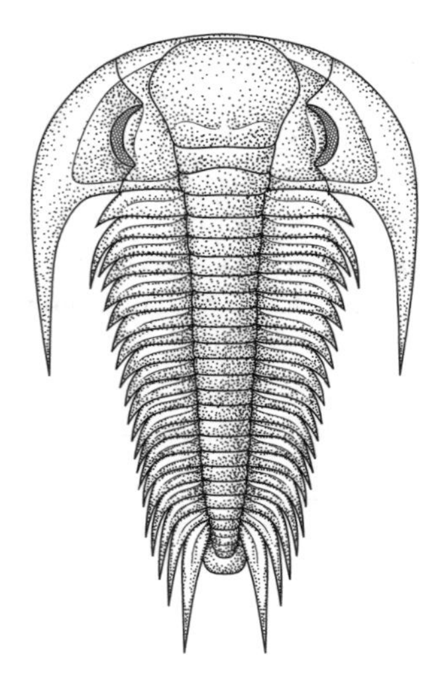

Middle Cambrian

An extremely large trilobite, growing up to 60 cm in length, with up to 21 segments in the thorax. Each segment ends in a long spine. The cephalon also bears a pair of long spines (called genal spines) pointing towards the pygidium. The many spines on this trilobite may have helped it to stay on top of the soft sediment, or may have had a protective purpose, as this species was not able to enroll completely.

Trinucleus

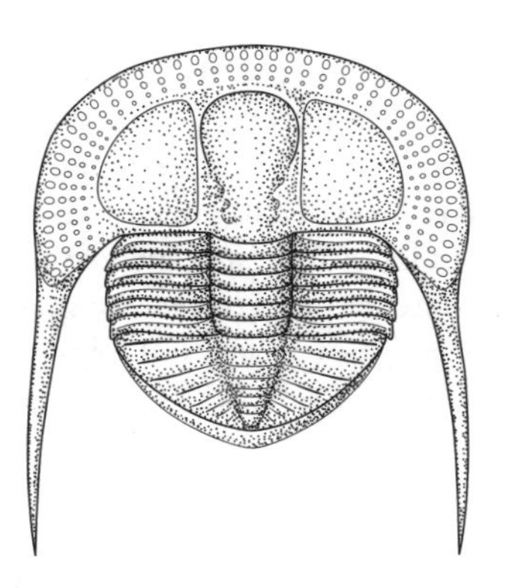

Ordovician

A blind trilobite with a wide fringe developed around the front of the cephalon. This likely had a sensory function. Genal spines extended from the fringe towards the rear of the animal. The glabella was inflated, suggesting that the stomach below would also have been large. In addition, the thorax would have been well above the sea bed in life. These two adaptations suggest that the species was a filter-feeder, suspending sediment in water below the body and then eating a mixture of mud and food which was processed in the large stomach. The thorax is relatively small, with six segments, and the pygidium is also small. Specimens tend to be small, 2–4 cm in length.

Agnostus

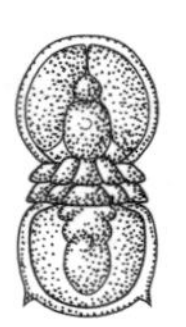

Cambrian–Ordovician

A tiny trilobite, usually less than 5 mm in length, with an unusual head and tail and only two thoracic segments. The species had no eyes or facial suture. Rare occurrences of this animal with preserved soft parts suggest that it was an active swimmer, which usually adopted a partly enrolled shape (see Fig. 8.2). It likely fed on material suspended above very soft sea floors and was typical of deep water assemblages.

Dalmanites

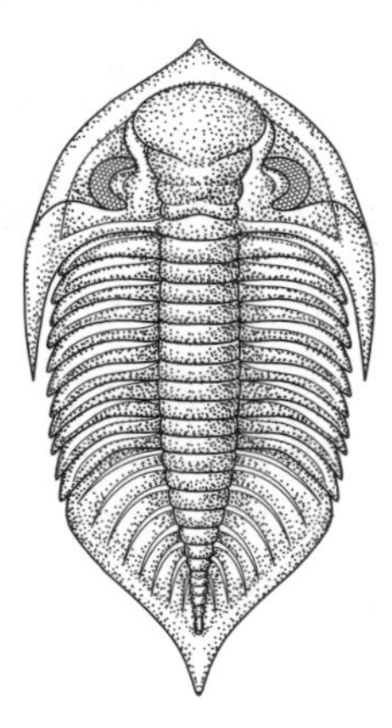

Silurian–Lower Devonian

An almond-shaped trilobite around 5 cm in length with a small spine extending forwards from the cephalon and another extending backwards from the pygidium. Large genal spines are developed on the cephalon, and are placed very close to the thorax. The thorax typically has 11 segments.

Calymene

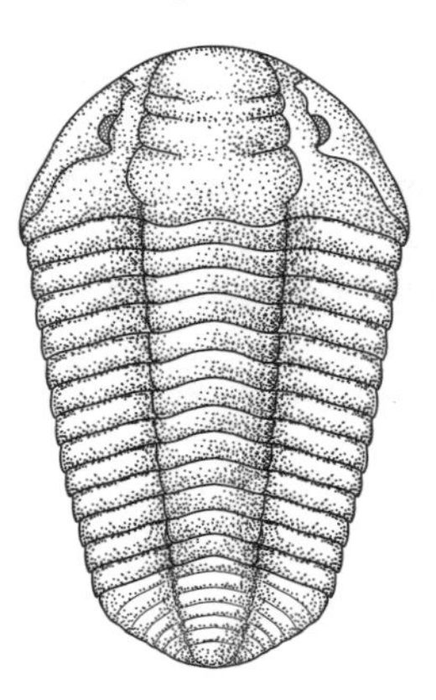

Silurian–Devonian

This is the trilobite most figured in this chapter. It is a small species, typically 2 cm in length. The cephalon is the widest part of the animal, and the thorax typically has 13 segments. The species was capable of complete enrollment, and is commonly preserved in this attitude. The fixed hypostome suggests that *Calymene* was predatory.

Cyclopyge

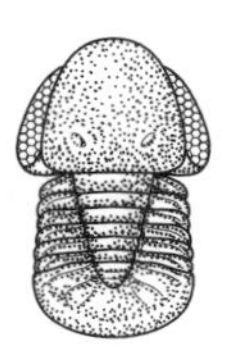

Ordovician

A large-eyed trilobite with a streamlined thorax and pygidium. The species was small, typically 4 cm in length, and is usually found in black shales. It had a wide geographic distribution. The features of this species are typical of pelagic trilobites, active swimmers who needed good all-round vision, including a view downwards.

Phillipsia

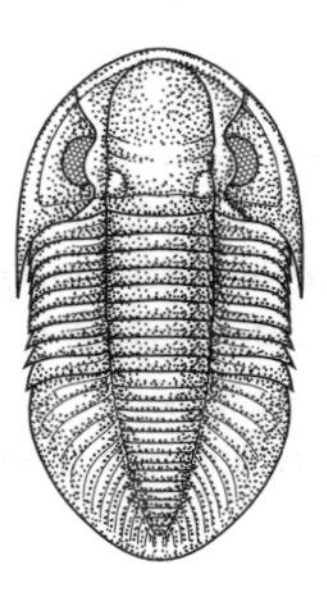

Lower Carboniferous

Another small trilobite, 2–4 cm in length. This trilobite had an oval shape, with a cephalon and pygidium of about the same size. The thorax between has nine segments. This is a member of an extremely long-lived and morphologically conservative group of trilobites, the proetids, which were amongst the last survivors of the group.

Deiphon

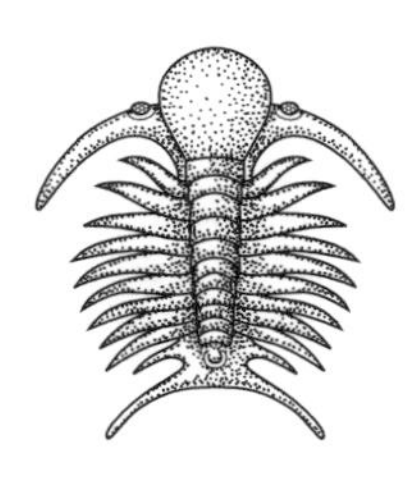

Silurian

A very spiny trilobite about 4 cm in length when fully grown. The cheeks were reduced and both cephalon and pygidium are remarkable for their pair of large, robust spines. The lifestyle of this unusual form is disputed. It may have been pelagic, and the eyes point forwards, but they are rather smaller than other known open water forms. It may well have been an active trilobite living on soft mud, where spines would have helped it stay on the sediment surface. This is known as a “snowshoe” adaptation.

Glossary

Axial lobe – central element of the thorax.
Cephalon – head of a trilobite.
Ecdysis – process of molting whereby most arthropods intermittently shed their external skeleton in order to grow.
Facial suture – break in the trilobite skeleton, usually around the eyes, that facilitated ecdysis.
Fixed cheek – part of the cephalon between the glabella and the facial suture; sometimes called the fixigena.
Free cheek – part of the cephalon outside the facial suture; sometimes called the librigena.
Glabella – raised part of the center of a trilobite head that protected the stomach.
Hypostome – lightly skeletonized cuticle that protected the mouth.
Pleural lobes – two lateral elements of the thorax.
Pygidium – trilobite tail.
Thorax – main body of a trilobite, where the segments are.

9 Molluscs

- Molluscs are divided into three main groups: the Gastropoda, Bivalvia, and Cephalopoda.
- Gastropods are the most diverse group of molluscs, living in terrestrial and all aquatic environments.
- The bivalve shell shape is constrained strongly by function so that mode of life can often be interpreted from shell morphology.
- Cephalopods are the most morphologically complex molluscs. As active predators, occupying the same ecological niche as fish, cephalopods may be considered to be the most sophisticated invertebrate group.

Introduction

Molluscs are an extremely diverse and abundant group. Most molluscs are marine, living at all levels from the intertidal zone to the abyssal depths. Some groups live in fresh water and others have adapted to live on land. The phylum includes animals with an external shell, for example, snails and oysters, as well as mainly soft-bodied forms, for example, slugs and squids.

Basic morphology

Most molluscs have an elongate, unsegmented body with a distinct head. The internal organs are held between a muscular foot, a modified lower part of the body, and a calcareous shell secreted by an underlying tissue known as the mantle (Fig. 9.1). The mantle tends to overhang the body forming a chamber at the posterior, the mantle cavity. This cavity contains the gills. The mouth opens anteriorly, at the other end of the mollusc. Sensory organs, such as eyes and tentacles, are concentrated in the head.

Shell morphology is extremely diverse and it performs a range of functions for different groups of molluscs. Shells may be coiled or straight, chambered or undivided, singular or two-valved. Primarily shells provide protection but they also may be used in burrowing or boring, or to enable buoyancy. The evolutionary development of some molluscs, such as octopus and squid, has tended to result in the loss of the shell. In other groups, such as the snails, the shell has become emphasized with time. Molluscs with a shell are more likely to fossilize and therefore this chapter will focus on such groups.

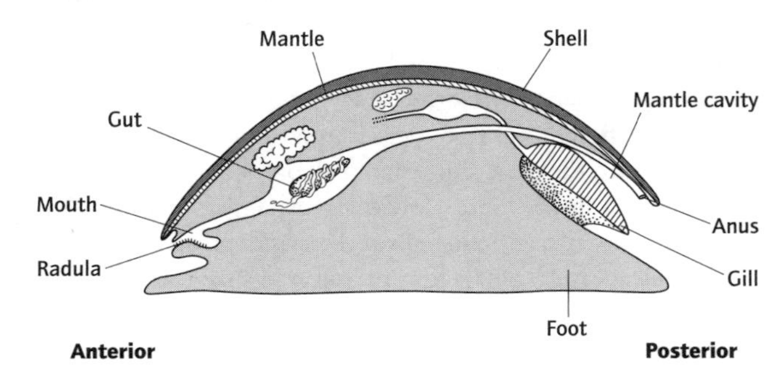

Fig. 9.1 Basic molluscan morphology.

Molluscan origins

Molluscs are first known from the early Cambrian. Most of these early forms did not persist for long and had an atypical appearance relative to that of modern members of the group. Molluscs probably evolved from animals similar to living flatworms.

Although distinct classes of molluscs have been established the relationships between the classes remain uncertain.

Classification

The nature of the shell is an important characteristic in mollusc classification. Gastropods, bivalves, and cephalopods are the dominant classes of fossil and living molluscs. A more detailed description of each class is given in Table 9.1.

Table 9.1 The main groups of molluscs.

Group	Description
Class Gastropoda	Single, undivided, coiled shell or shell-less. Mantle cavity faces the front. Muscular foot is flattened and used for locomotion. Head is well developed with eyes and tentacles, e.g., snails and slugs. Late Cambrian–Recent
Class Bivalvia	Two-valved, hinged shell encloses the body. No head but foot well developed. Gills are modified for respiration and filter-feeding, e.g., mussels, oysters, and scallops. Lower Cambrian–Recent
Class Cephalopoda	Internal or external shell or shell-less. Head and sensory organs well developed. Living representatives are intelligent and actively carnivorous. Late Cambrian–Recent
Subclass Nautiloidea	External, buoyant, chambered shell. Chambers are connected and partition walls are flat, e.g., *Nautilus.* Upper Cambrian–Recent
Subclass Ammonoidea	External, buoyant, chambered shell. Chambers are connected and partition walls are folded, e.g., ammonites. Lower Devonian–Upper Cretaceous
Subclass Coleoidea	Shell internal and reduced or absent, e.g., belemnites, squids, and octopuses. Carboniferous–Recent

Molluscan shell growth

Most molluscs have a spiral shell. Each group has evolved a small range of the possible spiral shapes that can be generated, adopting those which function best for their mode of life. Spiral shells can be described simply using four variables (Fig. 9.2). Understanding the underlying symmetry of apparently very different mollusc shells is important because it gives an insight into the basic similarity of all of these forms.

Gastropods usually have a low whorl expansion rate (low W), and a high rate of translation along the axis of growth (high T). Their shell aperture can have a complicated shape, and the rest of the shell is formed out of this pattern, like icing squeezed from a tube.

Ammonites sometimes had no translation rate at all, and formed planispirals. Ammonites with high expansion rates, and a small distance of the aperture from the axis (low D) are involute, folded up so that their later whorls hide the earlier ones; those with low expansion rates are evolute, with each whorl visible.

Bivalves have shells with very high whorl expansion rates (high W). This is necessary because the two valves have to be able to open. If bivalve shells coiled more the spires would interfere with one another and the shell would lock shut. However, bivalve shells do translate along the axis of coiling, which is why their plane of symmetry falls between the valves.

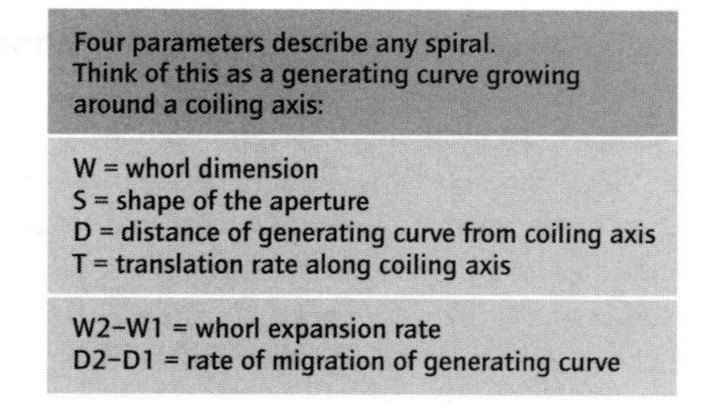

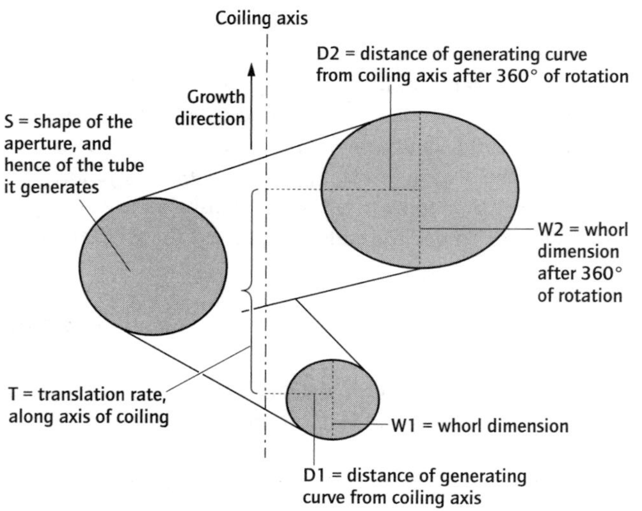

Fig. 9.2 How to describe mollusc shells as spirals.

Gastropods

Gastropods are the largest and most diverse class of molluscs. They live in marine, freshwater, and terrestrial environments and have exploited the widest variety of habitats and developed a remarkable range of feeding strategies. Gastropods first appeared in the early Cambrian and reached their peak diversity in the Cenozoic.

Gastropods may have a calcareous shell or be entirely soft bodied. They have a well-developed head and sensory organs and an expanded muscular foot (Fig. 9.3). In terrestrial gastropods the gills are lost and the mantle cavity is modified into an air-filled "lung". The defining characteristic of gastropods is torsion. At the embryonic stage, the visceral mass is rotated through 180°, bringing the mantle cavity to the front. In this position water currents flow more easily into the cavity. The shell coils to accommodate the gut which is more manageable when coiled.

There are three main subclasses of gastropods (Table 9.2). Although the classification is based on soft-part anatomy, mainly on the nature of the respiratory system, fossil gastropods are assigned to the different subclasses on the basis of shell shape.

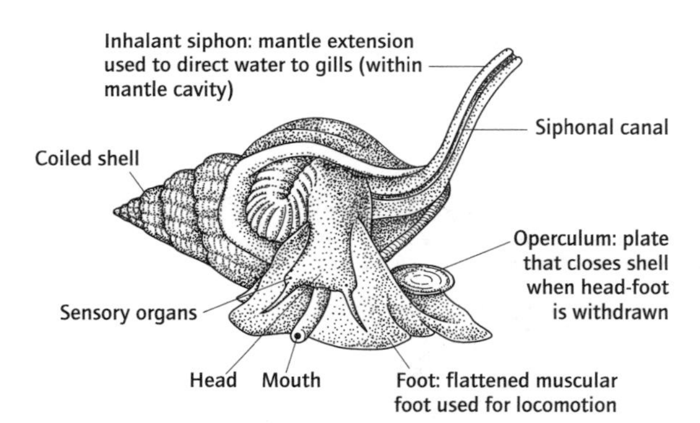

Fig. 9.3 General gastropod morphology.

Most fossil gastropods are prosobranchs and this subclass is divided into three orders:

1 Archaeogastropoda: Cambrian to Recent.
2 Mesogastropoda: Carboniferous to Recent.
3 Neogastropoda: Cretaceous to Recent.

These are not defined by a simple set of characteristics.

Ecology/evolutionary history

Cambrian gastropods were typically marine herbivorous grazers with low, coiled shells. By the Carboniferous, forms with a siphonal notch were common, indicating the presence of a siphon and therefore an infaunal mode of life. Palaeozoic gastropods generally occupied shallow water environments and were greatly affected by the end-Permian extinction.

During the Mesozoic prosobranchs diversified and deep-burrowing, long-siphoned prosobranchs originated in the Cretaceous. These forms dominate the gastropod fauna today. Carnivorous gastropods were important predators in the Cenozoic. The Tertiary also saw the radiation of gastropods into freshwater environments and the appearance of planktonic opisthobranchs. Radiation of the air-breathing pulmonates into the terrestrial environment began in the Jurassic. Today, gastropods are one of the most common groups of organisms living in the sea, in fresh water, and on land.

The aragonitic shell of most gastropods, which does not preserve well, and the high level of convergence shown by the group (that is, the large number of distantly related species that look alike because they share the same life habit) makes the evolution of fossil gastropods very difficult to elucidate.

Table 9.2 The main subclasses of gastropods.

Subclass	Respiratory system	Shell morphology	Habitat	Examples
Prosobranchiata Lower Cambrian–Recent	Gill in front	Cap shaped or conispiral	Mostly marine	Limpets, winkles, and whelks
Opisthobranchiata ?Carboniferous–Recent	Gill behind, due to detorsion	Shell lost or very reduced	Marine	Sea-slugs and sea hares
Pulmonata Mesozoic–Recent	Mantle cavity modified into "lung"	Where present, conispiral or planispiral	Terrestrial	Land snails and slugs

Bivalves

Bivalves are laterally compressed molluscs enclosed within a pair of hinged shells or valves. Valves are closed by the adductor muscles. The shell is opened by relaxing these muscles and water currents are drawn into the cavity. In the majority of bivalves the gill is modified for filter-feeding, although the earliest bivalves may have been deposit-feeders. Most forms are capable of limited movement and the foot can protrude into the sediment to enable them to burrow (Fig. 9.4). Bivalves are superficially similar to brachiopods, but have a very different internal anatomy and the shell is usually symmetric between the valves, rather than across the valve as in brachiopods.

Bivalves originated in the early Cambrian and have diversified, particularly after the end-Permian extinction, to exploit a wide range of aquatic environments. They are characteristic of Mesozoic and Cenozoic shallow water assemblages. Most bivalves are marine but there are some freshwater forms. As the shell forms of different species are adapted to the nature of the substrate, bivalves are very useful in palaeoenvironmental reconstruction.

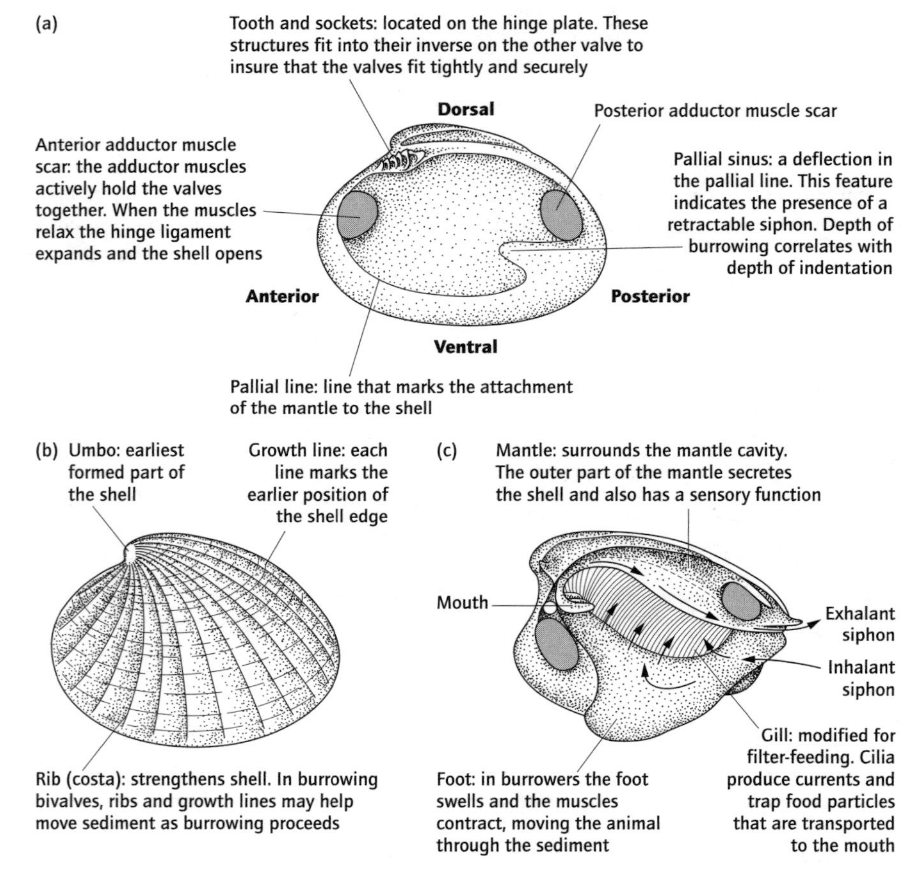

Fig. 9.4 Bivalve morphology: (a) general interior of a valve (right valve); (b) general exterior of a valve (left valve); and (c) internal morphology. Arrows show water currents in the mantle cavity.

Bivalve shells

Bivalve shells are multilayered. Two phases comprise the shell – an organic matrix and a crystalline calcareous component, in the form of aragonite or calcite ($CaCO_3$). Distinct shell structures have been identified that may be linked to the mode of life, for example, burrowing bivalves possess a shell structure that is resistant to abrasion. Environmental factors such as water temperature may also affect shell mineralogy.

The two halves of the bivalve shell are joined by a ligament that holds the valves in tension, allowing the shell to open when the adductor muscles relax. Teeth and sockets in the hinge area of the valves interlock to insure a tight fit when the shell is closed. Several patterns of dentition exist in bivalves, and this is a useful characteristic in classification. The main types are shown in Table 9.3.

Table 9.3 Types of bivalve dentition.

Dentition	Example
Taxodont Numerous teeth in radial or subparallel arrangement	*Arca*
Dysodont Small simple teeth at valve margin	*Mytilus*
Isodont Very large teeth positioned either side of ligament pit	*Pecten*
Schizodont Large grooved teeth	*Neotrigonia*
Heterodont Large teeth (termed cardinal teeth) flanked by smaller (lateral) teeth	*Venus*
Desmodont Teeth reduced to ridges or absent	*Lutraria*

Evolutionary history

The earliest bivalves are known from Lower Cambrian rocks. These primitive forms are extremely small and may be descended from an unusual group of primitive molluscs called rostroconchs. Although superficially similar to bivalves, rostrochonchs have a bilobed shell, without a functional hinge line. Strangely, no bivalves are recorded from the Middle and Upper Cambrian. A major period of diversification occurred in the early Ordovician. Groups arose with taxodont, dysodont, and heterodont hinges and a range of feeding strategies. Deposit-feeders, byssally attached bivalves, and burrowers colonized marginal and near-shore environments. After this rapid radiation the group stabilized and bivalves were not a particularly diverse or abundant group during the Palaeozoic. Nonmarine bivalves appeared in the Devonian and were abundant in the Carboniferous, particularly in deltaic environments.

During the early Mesozoic, bivalves underwent a second, more significant, radiation. The presence of a muscular foot and the development of siphons, by the fusion of the posterior edge of the mantle, gave bivalves an evolutionary advantage over brachiopods, allowing them to successfully exploit an infaunal mode of life. Predation pressure increased in the Mesozoic and bivalves were able to feed effectively from the secure position of a burrow within the sediment. By colonizing new infaunal environments, bivalves expanded into the intertidal zone, burrowed deeper into the sediment, and developed mechanisms for boring into hard substrates.

Diversification of epifaunal bivalves also occurred in the Mesozoic. Large, strongly inequivalved, recumbent forms developed. These are groups in which one valve is large and forms a body cavity for the whole animal, while the other valve is much smaller, often flattened, and functions as a lid. The most important of these groups are the rudist bivalves. They colonized carbonate shelves, adopting conical forms similar to corals or encrusting or lying on hard substrates. This group was short lived, originating in late Jurassic times and becoming extinct at the end of the Cretaceous.

Apart from the more bizarre groups, such as rudists, most bivalves survived the end-Cretaceous mass extinction. Bivalves have been ubiquitous in most shallow marine environments throughout the Cenozoic and they continue to be abundant in most present-day marine environments.

Bivalve ecology

Bivalves have adapted to live in a wide range of marine and freshwater environments. As shell shape is constrained by function, the mode of life of bivalves can be interpreted from the shell morphology.

The major life habits of bivalves are: (i) burrowing in soft substrates; (ii) boring and cavity dwelling; (iii) attached (cemented or by byssus threads); (iv) unattached recumbant; and (v) intermittant swimming.

Infaunal bivalves

Bivalves that burrow in soft substrates tend to be equivalved and have a distinct pallial sinus (Fig. 9.5). Burrowing is achieved by the foot, which penetrates the sediment and swells. The muscles in the foot then contract, drawing the shell down through the sediment. Circular, heavily sculptured bivalves tend to burrow more slowly than smooth elongate forms, although the nature of the sediment also affects the speed of burrowing. Most burrowing bivalves are siphonate. During burrowing the siphons are closed. Some infaunal species are deposit-feeders.

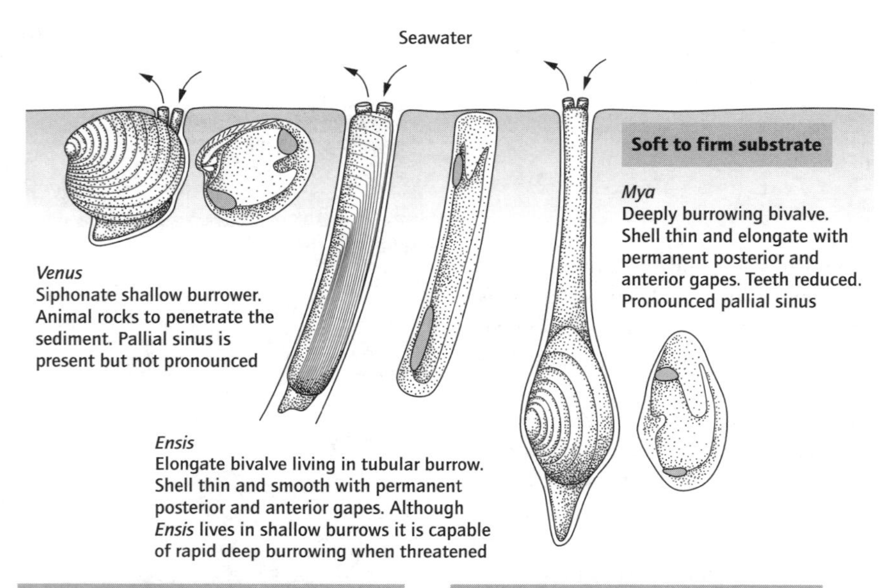

Shallow infaunal

Bivalves living in shallow burrows usually have thick, equivalved shells with an approximately circular outline. Adductor muscles are of the same size and there is usually a pallial sinus

Some shallow burrowers have pronounced ridges on the exterior of the shell, and this may enable the bivalves to "saw" through the sediment. It also helps to anchor the animal in the mud or sand where it is burrowing

Deep infaunal

Deep burrowing bivalves tend to have thin, elongate shells without surface sculpture. Some shells have permanent posterior and anterior gapes. Teeth often reduced. Pallial sinus pronounced

Fig. 9.5 Burrowing bivalves.

Boring bivalves

Bivalves that bore into hard substrates typically have elongate thin shells that are resistant to abrasion (Fig. 9.6). The shell edges are used to penetrate the material aided by acid secretions from the mantle. Some wood- or mud-boring bivalves have spines that scrape the substrate as the bivalve rotates. Some bivalves squat in borings made by other species. These bivalves grow to fit their cavity shape.

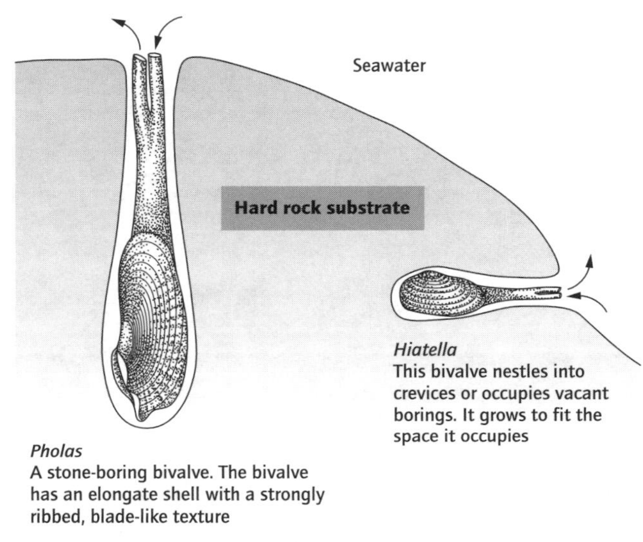

Fig. 9.6 Bivalves living in hard substrates.

Epifaunal and swimming bivalves

Epifaunal bivalves exploit three living strategies: (i) attachment to the substrate by byssus threads; (ii) cementation to hard surfaces; and (iii) recumbent, free lying on the sediment surface stabilized by the shell morphology (Fig. 9.7).

Byssally attached bivalves secrete threads of collagen that adhere to the substrate. Typically shells are elongate with a byssal gape, or notch. The anterior end of the shell and anterior muscle scar are usually reduced.

Cemented bivalves produce a calcareous fluid at the mantle margin that crystallizes, fixing the bivalve firmly to the substrate. Shells of cemented bivalves are usually inequivalved. Often there is only one large adductor muscle scar. Tropical cementers are usually covered in spines. As the bivalve grows it accommodates to the irregular surface on which it sits. This results in an irregular shell form that is characteristic of this life habit.

Recumbent, unattached bivalves have extremely asymmetric valves. Usually the lower valve is thickened, to increase stability, whilst the upper valve is reduced to a flat "lid".

Bivalves that are able to swim do so only intermittently. Shells of swimming valves tend to be thin, to reduce weight, although some forms have pronounced radial ribs. There is only one large adductor muscle scar. A single large muscle provides the strong contractions needed for swimming. Two pronounced "ears" are present, either side of the umbo, extending the hinge line. The umbonal angle in swimming bivalves is greater than that of byssally attached forms with a similar shape.

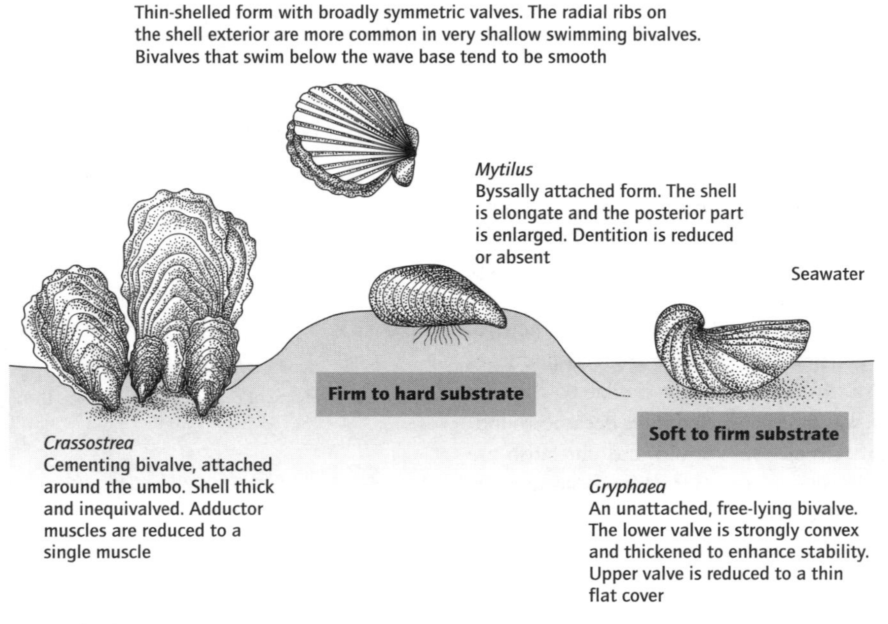

Fig. 9.7 Epifaunal and swimming bivalves.

Cephalopods

Cephalopods are the most morphologically complex group of molluscs. They occupy the same ecological niche as fish and they are arguably the most sophisticated group of invertebrates. The class includes active, jet-propelled predators with highly developed sensory structures. All cephalopods are marine.

The body of cephalopods is elongated so that the mantle cavity is anterior and the visceral mass is at the posterior end of the animal (Fig. 9.8). Living cephalopods swim using jet propulsion. Water is drawn into and expelled from the mantle cavity through the hyponome, a modified part of the foot. As the mantle cavity opens anteriorly the animal is propelled rapidly backwards. During slower locomotion the animal is able to direct its movement using the hyponome.

Cephalopods are divided into three subclasses: Nautiloidea, Ammonoidea, and Coleoidea. Nautiloids have an external, chambered shell with simple sutures between its chambers. Ammonoids also have an external shell, always coiled, with variable and more complicated sutures. Coleoids have an internal and reduced shell. In some coleoids the shell is absent.

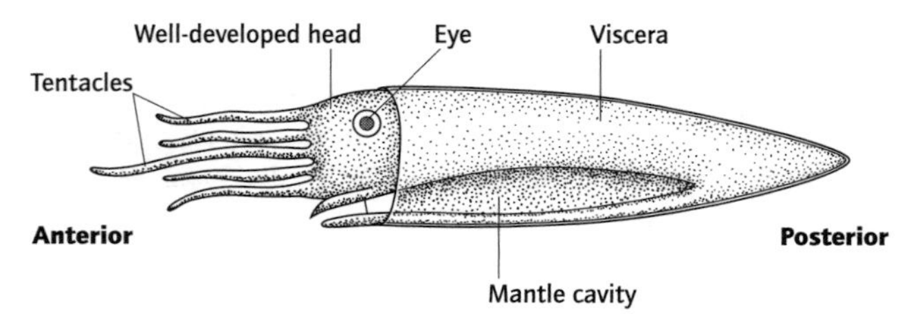

Fig. 9.8 General cephalopod morphology.

Evolutionary history

The first cephalopods were straight-shelled nautiloids. They appeared in late Cambrian times and underwent a rapid diversification in the Ordovician when they gave rise to the coiled forms that existed throughout the Palaeozoic and Mesozoic. Although never as abundant or diverse as ammonoids, they survive through to the present day (Fig. 9.9).

Ammonoids evolved from straight-shelled ancestors in the early Devonian. Their evolutionary history has been marked by a sequence of radiations followed by extinction. Peaking in diversity during the Jurassic, they declined through the remainder of the Mesozoic, becoming extinct at the end of the Cretaceous. Suture lines in ammonoids increased in complexity through time. Palaeozoic ammonoids generally had simple, straight suture lines whereas most Mesozoic ammonoids are characterized by complex sutures.

The history of coleoids is less well known due to their reduced, internal shell. The first true coleoids are recorded from Carboniferous rocks. Early coleoids were similar to the first nautiloids but the shell was internal. Belemnites became abundant in the Jurassic and Cretaceous. Squid and cuttlefish are known from the Jurassic, diversifying in the Cenozoic after the end-Cretaceous extinction event.

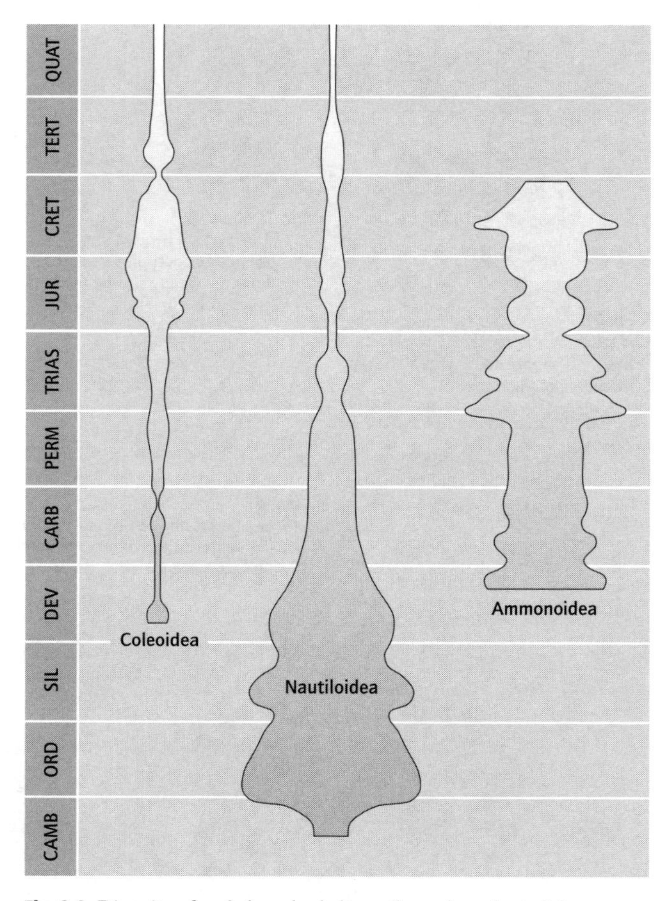

Fig. 9.9 Diversity of cephalopod subclasses through geological time.

Nautiloids

Nautiloids have a chambered, external shell that may be straight or coiled, with simple sutures. Most fossil nautiloids were active swimmers living close to the sea bed. The animal occupies the final shell chamber, the body chamber. The head, sensory functions, and hyponome are situated near the chamber opening and the visceral mass is at the rear. The animal is connected to the rest of the shell by the siphuncle, a tube extending from the body chamber to the protoconch (the first part of the shell to be formed) (Fig. 9.10).

Nautiloids with orthocone (straight) shells swam with their shell in a horizontal orientation. Shells were often modified in order to counterbalance the weight of the soft parts concentrated at the anterior end, thus maintaining stability. Nautiloids with coiled shells are more physically stable than straight-shelled forms. *Nautilus* is the only living cephalopod that retains an external, coiled shell. Living in cool waters in the southwest Pacific, at depths between 150 and 300 m, *Nautilus* relies on buoyancy control to adjust its position in the water column. *Nautilus* is an opportunistic feeder, grasping mainly crustaceans and small fish with its tentacles.

Nautilus is a poor swimmer. Seawater is drawn into the mantle cavity and expelled through the hyponome. As water is ejected a force is exerted on the shell, causing it to lurch forwards. When the mantle cavity is emptied the shell swings backwards generating a see-sawing motion. Consequently, *Nautilus* swims only for short distances and relies on buoyancy control to maintain its position in the water column during feeding. It rests during the day on the sea floor.

Buoyancy in *Nautilus*

Nautilus has an adjustable buoyancy mechanism that gives it neutral buoyancy at different depths of the water column. The shell chambers contain gas and seawater, the proportions of which can be changed (Fig. 9.11).

Initially the chambers contain seawater. The siphuncle removes ions from solution in the seawater, drawing water from the chambers into the mantle cavity. Gas bubbles then diffuse into the space making the animal more buoyant and able to float higher in the water column. By pumping ions into the chambers, *Nautilus* brings water back into the chambers, making the animal less buoyant so that it sinks to lower depths.

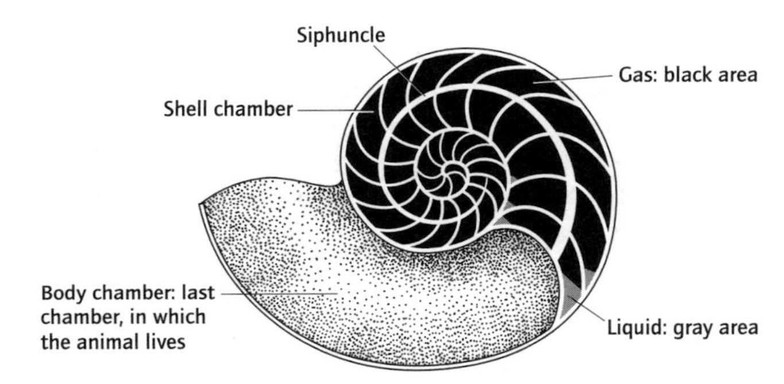

Fig. 9.11 The buoyancy mechanism of *Nautilus*.

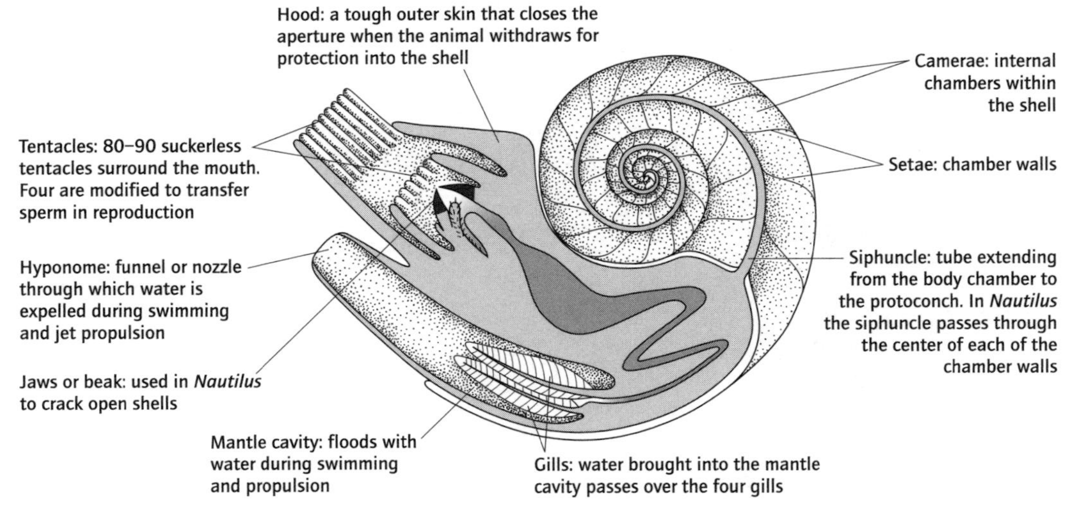

Fig. 9.10 Morphology of *Nautilus*.

Ammonoids

Most ammonoids had a chambered, planispirally coiled shell with complex sutures. As with nautiloids, the chambers were connected by a tube called the siphuncle, although in ammonoids the siphuncle usually ran along the outer, ventral margin of the shell rather than through the center of the chambers. The buoyancy mechanism of ammonoids was therefore similar to that of nautiloids.

Ammonoid suture

Sutures are the lines marking the junction between the chamber wall and the ammonoid shell. The suture pattern is an important feature used in ammonoid classification. Only ammonoids with an ammonitic suture pattern should be termed ammonites. Sutures are described in terms of saddles and lobes. Saddles are curved sections of line that "point" towards the body chamber. Lobes are the converse – curves that are directed away from the body chamber. The complexity of the suture pattern increased through time. Early sutures were gently sinuous whilst the sutures of Cretaecous ammonoids had an intricate, fern-like form (Table 9.4).

Evolutionary history

Ammonoids first appeared in the early Devonian. They evolved from straight-shelled nautiloids and the early forms had simple, straight sutures. Ammonoids rapidly diversified in the Devonian. By late Devonian times goniatitic sutures had developed as setae (chamber walls) became folded. Goniatitids were the dominant Palaeozoic ammonoid form. In the Carboniferous more complicated, ceratitic, suture patterns developed. Ammonoids persisted through the Palaeozoic but suffered a major crisis at the end of the Permian.

Few ammonoids survived into the Triassic, but those that did radiated rapidly. Triassic ammonoids had ceratitic sutures, although complex ammonitic sutures developed towards the end of the period. More intricate sutures may have increased the strength of the shell, enabling ammonoids to live in deeper water. It has also been suggested that the greater surface area of the folded wall would have increased the speed with which ions were removed from solution and hence increased the speed with which the animal could change its density.

The end-Triassic extinction eliminated most of the ceratitic ammonoids. Following the extinction, ammonoids rediversified, this time with the more complicated ammonitic suture. These are the ammonites, whose diversity peaked in the Jurassic. Towards the end of the Cretaceous diversity began to diminish and none survived into the Tertiary.

All groups of ammonoids are of use in biostratigraphy, because of their wide distribution (enhanced by their ability to float after death) and their rapid evolution. Goniatitids are extremely useful in Carboniferous sediments, ceratitids in the Triassic, and ammonites through the rest of the Mesozoic.

Table 9.4 Ammonoid suture patterns. Arrows point to the aperture. L, lobe; S, saddle.

Suture type	Description
Ammonitic Permian to Cretaceous; dominant in the Jurassic and Cretaceous	Complex sutures. Lobes and saddles crenulated
Ceratitic Carboniferous to Triassic	Saddles undivided, land lodes crenulated
Goniatitic Upper Devonian to Upper Permian	Usually eight simple lodes and saddles
Orthoceratitic Lower Cambrian to Recent	No lobes or saddles

Ammonite morphology

Ammonites had an external, chambered, usually planispiral shell. The shell can be divided into three parts: the body chamber, where the animal lived; the phragmocone, the chambered part of the shell (each chamber represents part of a previous body chamber); and the protoconch, the first chamber to form (Fig. 9.12). Chambers are connected by a siphuncle positioned along the outer margin, or venter, of the shell. Shell morphology is variable (Fig. 9.13) and the external surface is often heavily sculptured with spines, tubercles, and ribs.

Sexual dimorphism

Mature ammonite shells collected from the same horizon can often be divided, on the basis of size, into two distinct morphological groups. The smaller ammonites are referred to as microconchs and the larger type as macroconchs (Fig. 9.14). Microconchs may also have a modified aperture with lateral extensions, the lappets. The function of the lappets is unknown but it may be linked to sexual reproduction. Although microconchs and macroconchs may simply be closely related species, new characters appear in both groups simultaneously suggesting that they are males and females of the same species, though which is which is unknown.

Heteromorphs

Some ammonite groups developed bizarre or "heteromorph" shell forms (Fig. 9.15), particularly in the late Cretaceous. Originally these forms were considered nonfunctional, evolutionary dead ends. However, physical modeling has shown that they were stable and well adapted for floating within the water column. Furthermore, it has been shown that heteromorphs gave rise to more conventional ammonite groups.

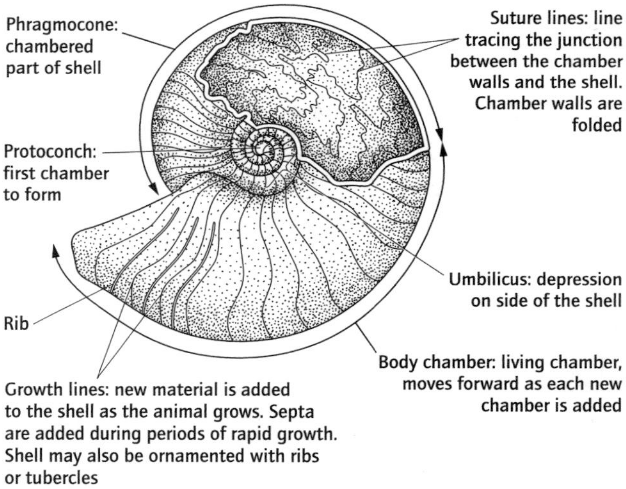

Phragmocone: chambered part of shell

Suture lines: line tracing the junction between the chamber walls and the shell. Chamber walls are folded

Protoconch: first chamber to form

Umbilicus: depression on side of the shell

Rib

Body chamber: living chamber, moves forward as each new chamber is added

Growth lines: new material is added to the shell as the animal grows. Septa are added during periods of rapid growth. Shell may also be ornamented with ribs or tubercles

Fig. 9.12 Ammonite hard-part morphology.

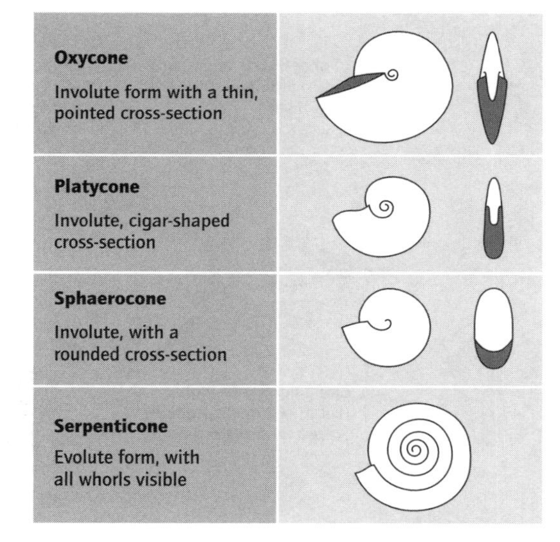

Fig. 9.13 Shell shape terminology.

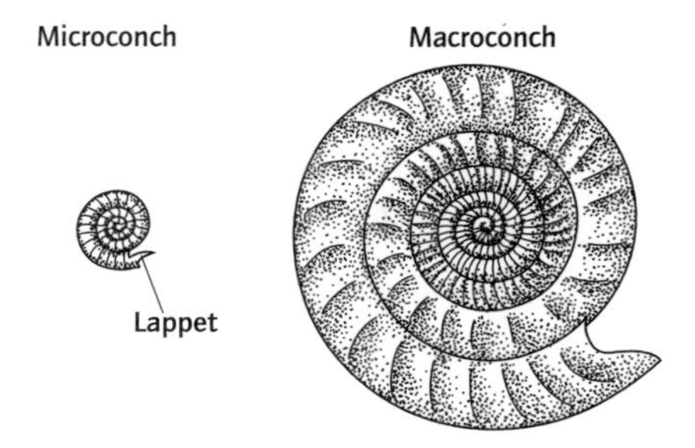

Fig. 9.14 Sexual dimorphism in *Perisphinctes* (× 0.2).

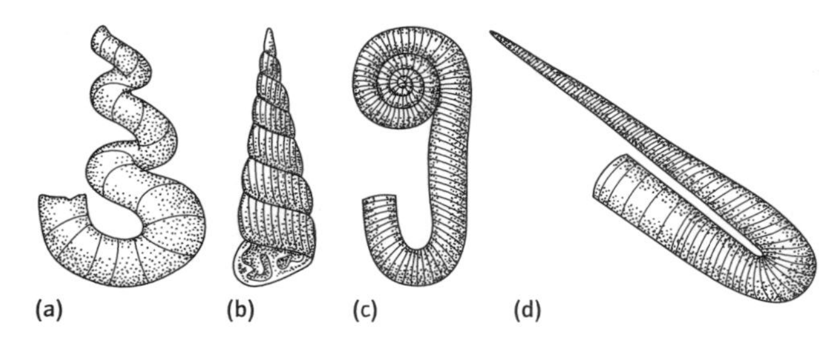

Fig. 9.15 Ammonoid heteromorphs, in proposed life orientation: (a) *Hyphantoceras*; (b) *Ostlingoceras*; (c) *Macroscaphites*; and (d) *Hamulina*.

Palaeoecology

Interpretations of the ammonite mode of life are based on the behavior of *Nautilus*. However, ammonite shell morphology is extremely diverse. The effect of shell shape on lifestyle has been investigated using numerical modeling to assess the stability of the shell within the water column and by observing the hydrodynamic properties of different shells in the laboratory.

Any body immersed or floating in a liquid is acted on by a buoyant force equal to the weight of liquid displaced. The point through which this force acts is the center of buoyancy. The point through which the resultant force of gravity passes is the center of gravity. In a stable orientation the center of gravity lies directly below the center of buoyancy. The greater the distance between the centers, the greater the stability. By estimating the volume and weight of ammonites during life, the center of gravity and buoyancy of the animal can be calculated (Fig. 9.16). This work has established the probable orientation of ammonites with differing shell morphologies within the water column, helping to identify their likely mode of life (Fig. 9.17).

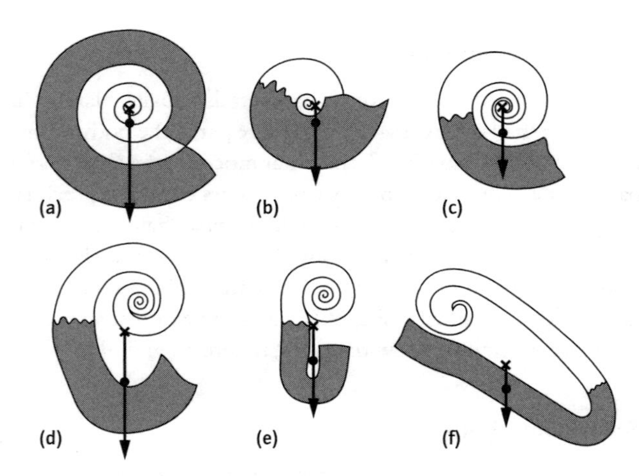

Fig. 9.16 Life attitudes of ammonites. The arrows point from the center of buoyancy to the center of gravity. The solid dot is the center of buoyancy, and the cross is the center of gravity. The shaded area is the body chamber: (a) *Dactylioceras*; (b) *Ludwigia*; (c) *Crioceras*; (d) another species of *Crioceras*; (e) *Macroscaphites*; and (f) *Lytocrioceras*.

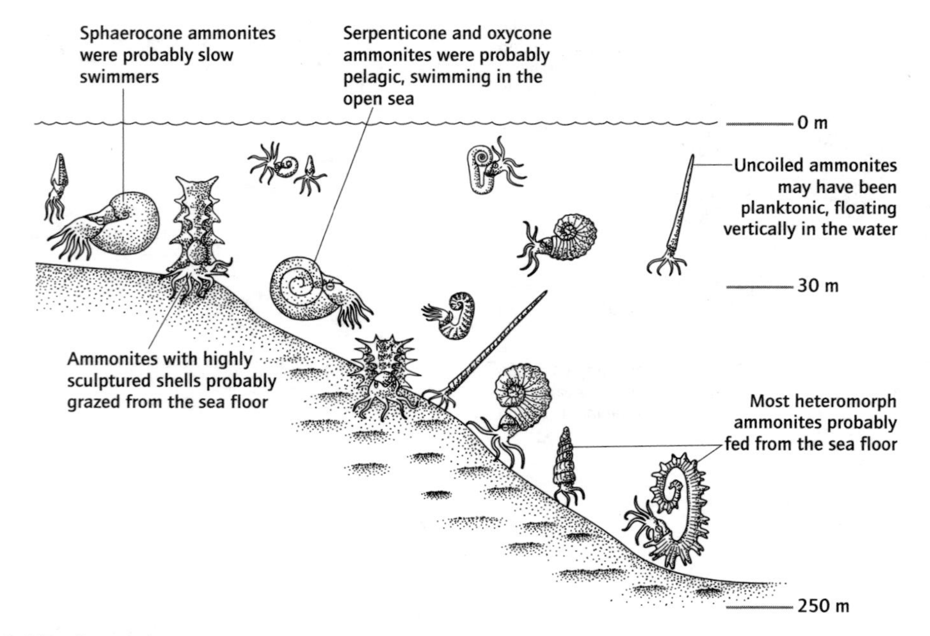

Fig. 9.17 Possible life habits of ammonites.

Coleoids

The subclass Coleoidea includes all living cephalopods except *Nautilus*. The shell in coleoids is internal and reduced or even absent. Modern representatives are diverse, highly modified, active predators with sophisticated sensory systems. Cuttlefish have an internal shell with a buoyancy function. Squids are streamlined swimmers with an internal cartilaginous rod (Fig. 9.18). Although they do not have a chambered shell they may increase their buoyancy through the retention of ammonia, which is a metabolic waste product less dense than seawater. Octopuses are benthic, shell-less coleoids with "webbed" arms that allow the animal to drift with the currents. Belemnites make up the majority of the fossil coleoids. Characterized by an internal skeleton with a robust, bullet-shaped, calcite counterweight, belemnites are abundant in Jurassic and Cretaceous rocks.

Belemnites

Belemnites had an internal skeleton unlike any living coleoid. It can be divided into three parts: (i) the robust anterior counterweight – the rostrum or guard; (ii) the buoyancy mechanism – the phragmocone, a chambered conical section with a siphuncle; and (iii) the pro-ostracum – the support for an open body chamber. The rostrum is made from solid calcite composed of radially arranged, needle-like crystals (Fig. 9.19).

Belemnite soft tissues are known from a few exceptionally preserved specimens. The tentacles had small hooks and some belemnites had ink sacs similar to those of modern squids.

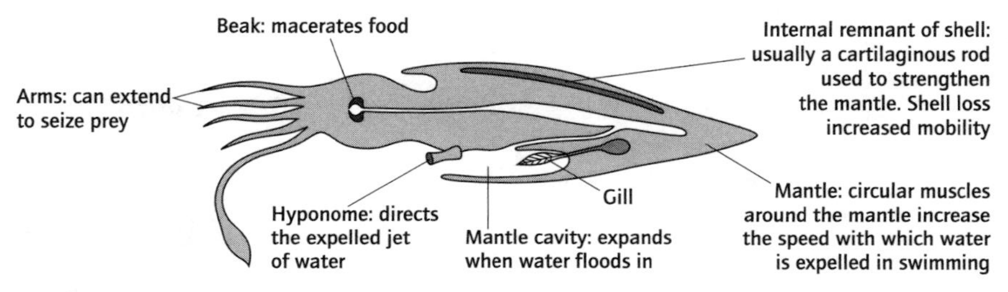

Fig. 9.18 Squid soft-part morphology.

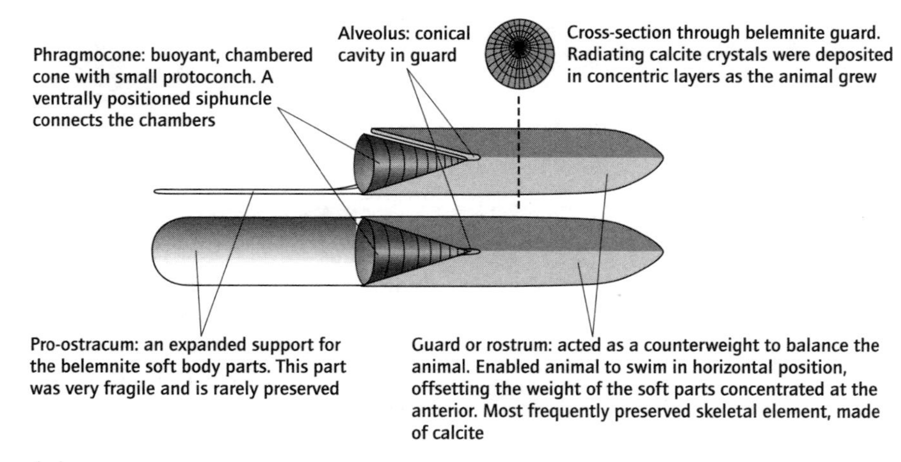

Fig. 9.19 Belemnite morphology.

Mya

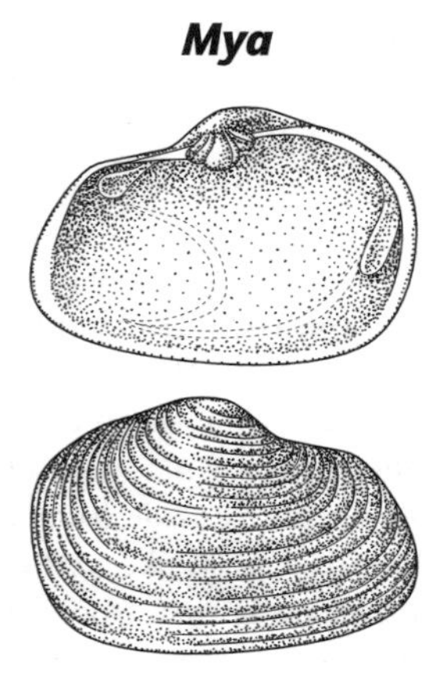

Bivalve

Oligocene–Recent
This bivalve has an elongate, smooth shell with a posterior gape. The shell is approximately 8 cm from umbo to shell edge. Dentition is absent and there is a deep pallial sinus.

Modern species are infaunal, living in soft sediment in burrows 30 cm deep. The long siphons extend to the seawater. Both siphons are enclosed within a protective sheath.

Ensis

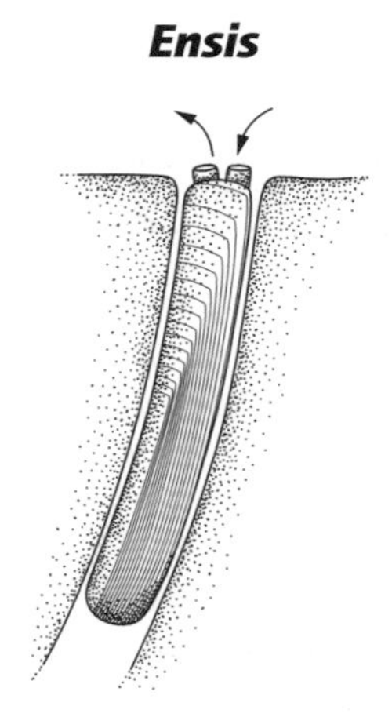

Bivalve

Eocene–Recent
Characterized by an extremely elongated, thin, featureless shell (approximately 12 cm in length) with both posterior and anterior gapes. *Ensis* lives infaunally in muds and sands in the intertidal zone. During feeding, the anterior part of the shell is close to the sediment–water interface. During low tide the animals burrows actively down into deeper sediments using the muscular foot.

Teredo

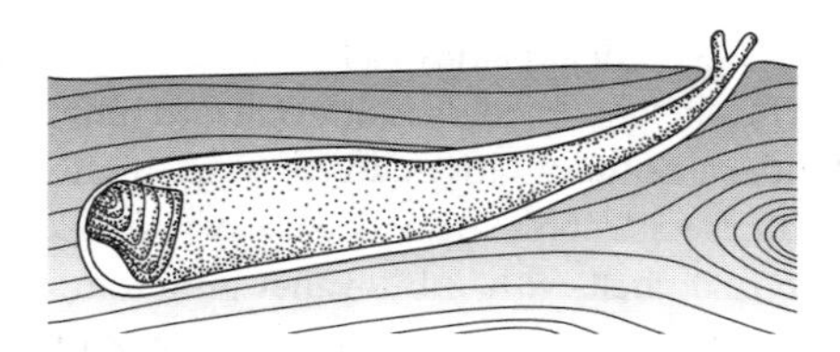

Bivalve

Eocene–Recent
Teredo is a highly specialized bivalve able to bore into wood. The cylindrical shell is extremely reduced (approximately 1.2 cm from umbo to shell edge). Sharp sculpture on the external surface is used for tunneling into the substrate. The shell is reduced and the animal is essentially worm-like and lives encased in the burrow, growing to fill the excavated space.

Radiolites

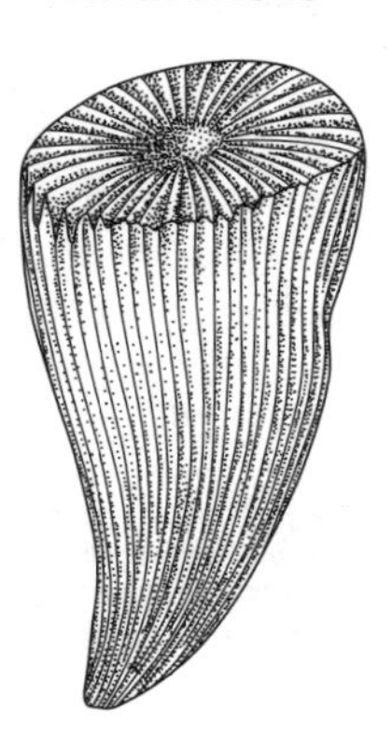

Bivalve

Cretaceous
Known as rudist bivalves, such highly modified forms of cementing bivalves were common in the Cretaceous. *Radiolites* has two strikingly different valves. The lower valve is conical (approximately 12 cm in height) with thick walls and the upper valve is reduced to a small flat lid. Such coral-like rudists tended to grow in goups, feeding in calm, clear waters above the sea bed.

Turritella

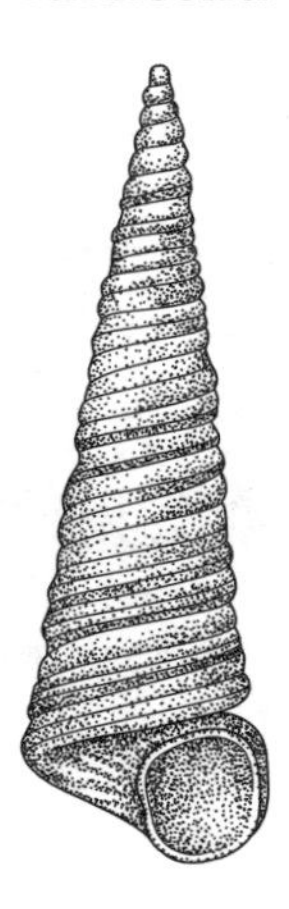

Gastropod

Cretaceous–Recent
Turritella belongs to the subclass Prosobranchiata, order Mesogastropoda. The multiwhorled shell with a high pointed spire has a simple, circular aperture and spiral ribbing on the external surface. There is no siphonal canal. The height is approximately 5 cm.

Modern species are usually found buried with the spire facing downwards in soft sediment in marine, shallow water environments. Food is collected from water drawn into the mantle cavity.

Planorbis

Gastropod

Jurassic–Recent
Belonging to the subclass Prosobranchiata, order Mesogastropoda, this freshwater gastropod has an almost planispiral shell (diameter approximately 1 cm). Although the morphology varies within the genera most species have smooth shells.

Living in a range of freshwater environments, *Planorbis* feeds on algae and plants. Some species live entirely within water whilst others need to surface for air.

Buccinum

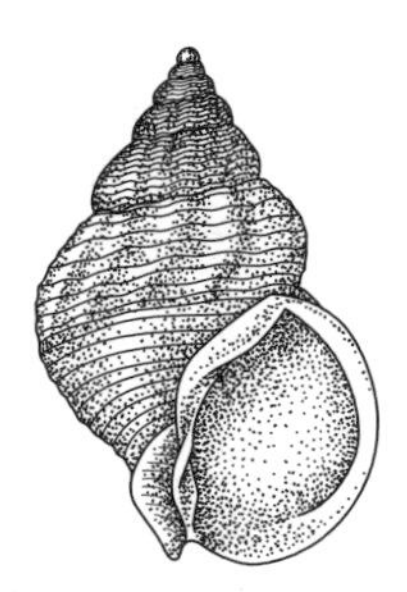

Gastropod

Pliocene–Recent
This mesogastropod has a moderately high-spired shell with an oval aperture and short siphonal canal. Typically the shell has an external, ribbed sculpture. The height is approximately 8 cm.

The visceral mass is situated in the shell and is helically coiled. For protection the head and foot can be withdrawn into the shell and the operculum closes the aperture. Living in seawater up to a depth of 200 m, *Buccinum* lives semi-infaunally with its siphon extended above the sediment drawing clean water into the mantle cavity. *Buccinum* is carnivorous.

Hygromia

Gastropod

Eocene–Recent
Hygromia is a terrestrial gastropod, the bristly snail, belonging to the subclass Opisthobranchiata. It has a modified, untwisted, mantle cavity that functions as an air-breathing lung. The shell is generally thin, smooth, and conispiral. Its height is approximately 5 mm.

Hygromia is found in a range of terrestrial habitats and is most common in humid environments with calcium-rich soils.

Patella

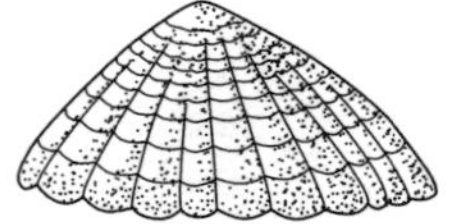

Gastropod

Eocene–Recent

This archeogastropod is characterized by an uncoiled, conical, cap-like shell. Pronounced ribs that radiate from the apex strengthen the shell. The height of the cone is approximately 3 cm.

Patella lives in the intertidal zone and clings to rocks using its foot. During low tide the animal "clamps down" to avoid desiccation. At high tide the gastropod grazes the rock surface for encrusting algae.

Gastrioceras

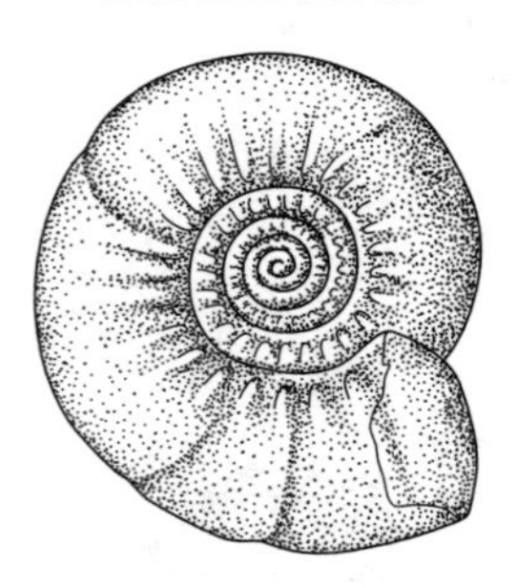

Cephalopod

Upper Carboniferous

This Palaeozoic ammonoid has a goniatitic suture. The shell is inflated (cadicone) and the external surface is finely ribbed. The shell diameter is approximately 5 cm. Small tubercles line the margin of the moderately deep umbilicus.

Gastrioceras is found in marine shales and is a useful zone fossil for the Upper Carboniferous.

Dactylioceras

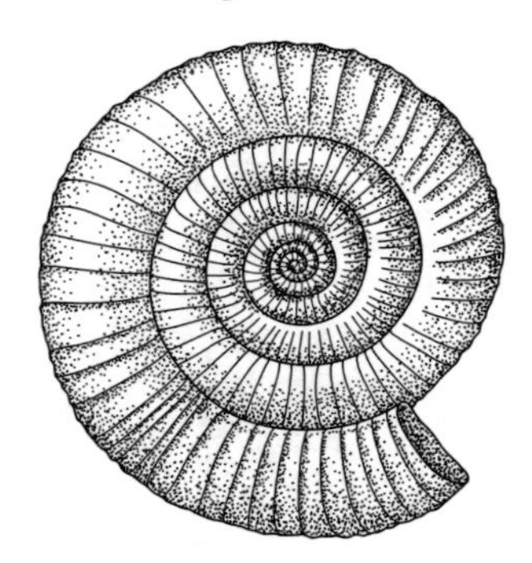

Cephalopod

Lower Jurassic

This ammonite is typically serpenticone and the external surface is ribbed (the shell diameter is approximately 6.5 cm). The body cavity is tubular and elongated.

Numerical modeling has shown that the center of gravity and the center of buoyancy are very close together (see Fig. 9.16). This means that the ammonite could have adopted a number of floating orientations within the water column but would have lacked stability.

Amaltheus

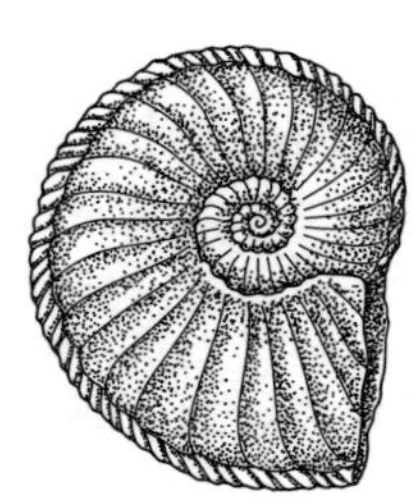

Cephalopod

Lower Jurassic

Amaltheus has a compressed, oxycone shell sculpted with curved, sinuous ribs. A "pie crust" keel is developed along the venter. The suture pattern is ammonitic. The shell diameter is approximately 8 cm.

The body chamber is large and positions of the centers of gravity and buoyancy show that the animal floated in a position similar to that adopted by modern *Nautilus*.

Hildoceras

Cephalopod

Lower Jurassic
This ammonite has an evolute shell with a wide umbilicus. The whorl cross-section is broadly quadrilateral. The external surface has a distinctive sickle-shaped ribbing pattern. The venter has a sharp keel flanked by a depression, or sulcus, on each side. The suture pattern is ammonitic. The shell diameter is approximately 6 cm.

Kosmoceras

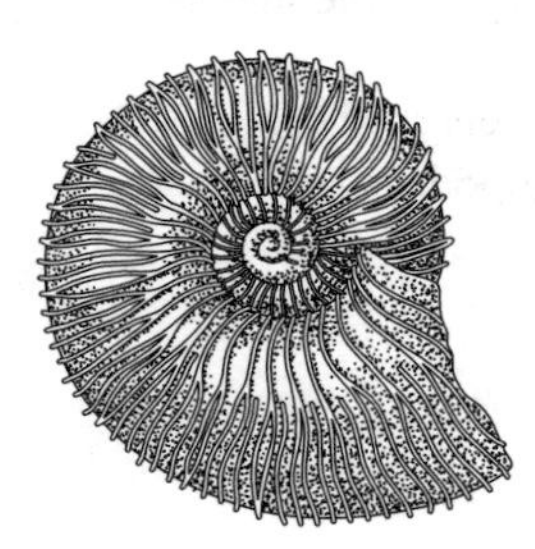

Cephalopod

Middle Jurassic
This ammonite has a compressed shell with pronounced ribs that bifurcate towards the venter. The suture pattern is ammonitic. Sexual dimorphism is known. In the microconch, the smaller dimorph, the aperture is compressed and the lappets are developed. The shell diameter of the macroconch is approximately 5 cm.

Scaphites

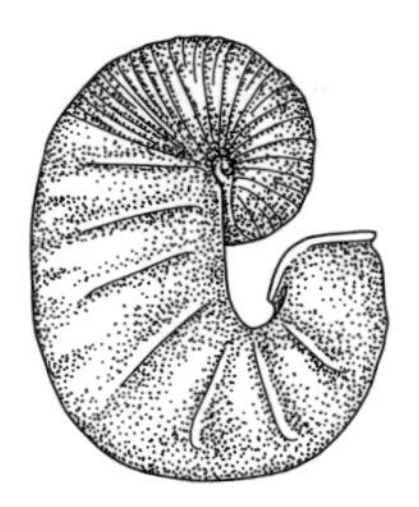

Cephalopod

Cretaceous
Scaphites is a heteromorphic ammonite typical of the Cretaceous period. This partially uncoiled ammonite has a body chamber in the form of a hook with a slightly constricted aperture that faces upwards. The external surface of the shell is ribbed. The "height" of the shell was typically 7 cm.

Due to the orientation of the aperture it is thought that *Scaphites* floated passively near the sea surface and that its swimming ability was limited.

Neohibolites

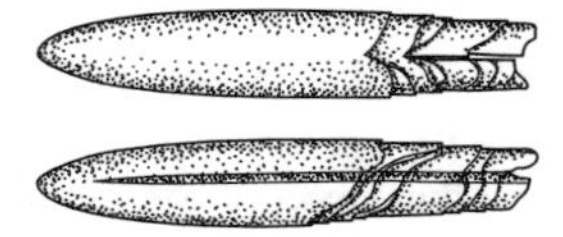

Cephalopod

Lower Cretaceous
This belemnite has a small, spindle-shaped guard (approximately 4 cm in length) with a long ventral groove in the area around the alveolus.

The soft-body morphology of belemnites is known from exceptionally preserved individuals associated with fossil Lagarstätten. Such specimens have long hooked tentacles and ink sacs.

Glossary

Adductor muscles – posterior and anterior muscles that hold the valves together in bivalves.
Alveolus – conical cavity in belemnite guard.
Byssal notch – gap through which byssal threads pass in bivalves.
Camera – chamber in cephalopod shell.
Equivalve – refers to bivalves with valves of similar size and shape, such as cockles.
Guard – bullet-shaped, solid calcite counterbalance found in belemnites.
Heteromorphs – aberrant forms of the ammonoid shell.
Hood – outer skin in nautiloids that closes the aperture when the animal withdraws into the shell.
Hyponome – funnel through which water is expelled during jet propulsion in cephalopods.
Inequivalve – refers to bivalves with differently sized and shaped valves, such as *Gryphaea*.
Lappet – extension to the ammonite aperture found in microconchs.
Macroconch – larger ammonite sexual dimorph.
Mantle – tissue that secretes and underlies the shell.
Mantle cavity – space formed by the mantle overhanging the mollusc's body. Contains the gills.
Microconch – smaller ammonite sexual dimorph.
Operculum – plates closing the aperture when the gastropod withdraws into the shell.
Pallial line – line marking where the mantle attaches to the shell.
Pallial sinus – deflection of the pallial line marking the presence of a retractable siphon.
Phragmocone – chambered part of the cephalopod shell.
Pro-ostracum – expanded, spatular-like support for belemnite soft tissues.
Protoconch – first chamber to form in ammonoid shell.
Radula – rasping tongue unique to molluscs.
Septa – chamber wall in a cephalopod shell.
Siphonal canal – gutter-like extension in the gastropod shell used to support the siphons.
Siphons – tube-like extensions of the mantle developed in gastropods and bivalves.
Siphuncle – tube extending from the body chamber to the protoconch that removes ions from solution and enables the proportion of gas and seawater in the chambers of the cephalopod shell to be altered.
Suture – in gastropods this is the line marking the junction of two whorls. In cephalopods it is the line tracing the junction of the chamber wall with the shell.
Umbo – first part of the bivalve shell to form.
Venter – outer margin of the shell in ammonoids.
Viscera – internal organs.
Whorl – one complete shell coil of 360°. Term is used for gastropod and cephalopod shells.

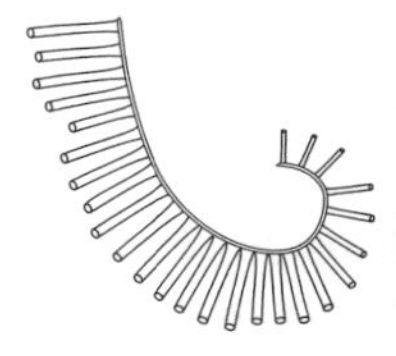

10 Graptolites

- Graptolites included planktonic and benthic species.
- The planktonic graptoloids were the most abundant Lower Palaeozoic zooplankton.
- Their wide distribution and rapid evolution make them extremely useful for biostratigraphy.
- They were an enigmatic group, but with living relatives called pterobranchs.

Introduction

Graptolites are an extinct group of marine, colonial animals that built a skeleton from a variety of proteins dominated by collagen. They varied in colony size from 2 mm to over 1 m in length, and in form from simple sticks to complicated bushy shapes. They belong to an obscure phylum, the Hemichordata, which has only a handful of living members. Of these, the pterobranchs are thought to be the closest relatives of graptolites.

The oldest graptolites are benthic species from the Middle Cambrian. Benthic groups diversified in the Ordovician, and a variety of poorly understood forms thrived at this time, including the tuboids, cystoids, camaroids, and crustoids. The most common benthic graptolites were the bushy dendroids, and these also had the longest range of any graptolite, surviving until the late Carboniferous. Benthic graptolite species were widespread but rarely common. They form a small component of most shallow water fossil assemblages from the Palaeozoic.

However, in the early Ordovician this benthic group gave rise to the planktonic graptoloids, which had a much more spectacular history. These organisms were the first really abundant and easily preserved macrozooplankton. They radiated quickly into a wide variety of extremely regular shapes and spread throughout the oceans of the world.

Initially, most graptoloids had many branches, but the number of these was reduced consistently until, by late Ordovician times, the fauna was dominated by species with only two branches arranged back-to-back. The biggest extinction survived by graptolites, at the end of the Ordovician, led to a major change in their form, with Silurian and Devonian assemblages being dominated by one-branched graptolites. Although detailed graptolite biostratigraphy is the province of real specialists, little expertise is needed to distinguish between Ordovician and Silurian faunas, and between the early and later parts of each period (see Table 10.1).

The greatest number of graptolite species was achieved in the early Ordovician, and successive crises reduced diversity; after each of which the recovery generated a smaller number of species. Graptoloids became extinct in the mid-Devonian.

The rapid speciation and wide distribution of graptoloids has made them pre-eminent in Lower Palaeozoic biostratigraphy. Although they are most common in offshore assemblages, especially in black shales, they are found in shallow water often enough to be generally useful. In parts of the Lower Silurian they allow the construction of a relative timescale that can divide this period into units of less than half a million years.

They are believed to have been filter-feeders, but no detailed soft parts have been found and inferences about their mode of life are made on the basis of hard-part morphology, species distribution, and studies of their living relatives.

Morphology

Graptolites had a relatively simple construction, and this is especially true for the planktonic graptoloids.

Benthic species built branches, or stipes, made of several types of cups, or thecae, in which the zooids lived. These thecae stick out from the stipe at regular intervals, and when a stipe is sectioned a bundle of thecae can be seen. The colony, or rhabdosome, of the graptolite is constructed from a set of stipes, arranged in characteristic patterns. In benthic graptolites this pattern is often irregular in detail, and the most common form is of a wide cone, with the stipes held together by a set of binding dissepiments. The rhabdosome was fastened to the sea bed in life with a holdfast developed from the sicula, a conical theca that was the first part of the colony to skeletonize (Fig. 10.1a).

In planktonic graptoloids the holdfast is replaced by a nema or virgula extending from the top of the sicula. This feature is usually a simple thread, but can have vanes or other additions. Stipes were built out from the sicula in extremely regular patterns, and are made from only one type of theca per species (Fig. 10.1d).

In detail, the construction of the colony was in two parts. First, the main shape was built from a series of half rings of collagen plastered on to the thecal apertures. Then, a second phase of consolidation and strengthening involved the addition of collagen bandages to both sides of the thecal wall (Fig. 10.1b).

A detailed terminology has developed for describing the number and attitude of stipes on a colony. The most important elements of this scheme are shown in Fig. 10.1c.

It is inferred that there was a soft-part connection between all of the zooids in the colony, and that one zooid inhabited a single theca. The shape of the zooids is unknown in detail, but is thought to have been similar to the appearance of modern pterobranchs.

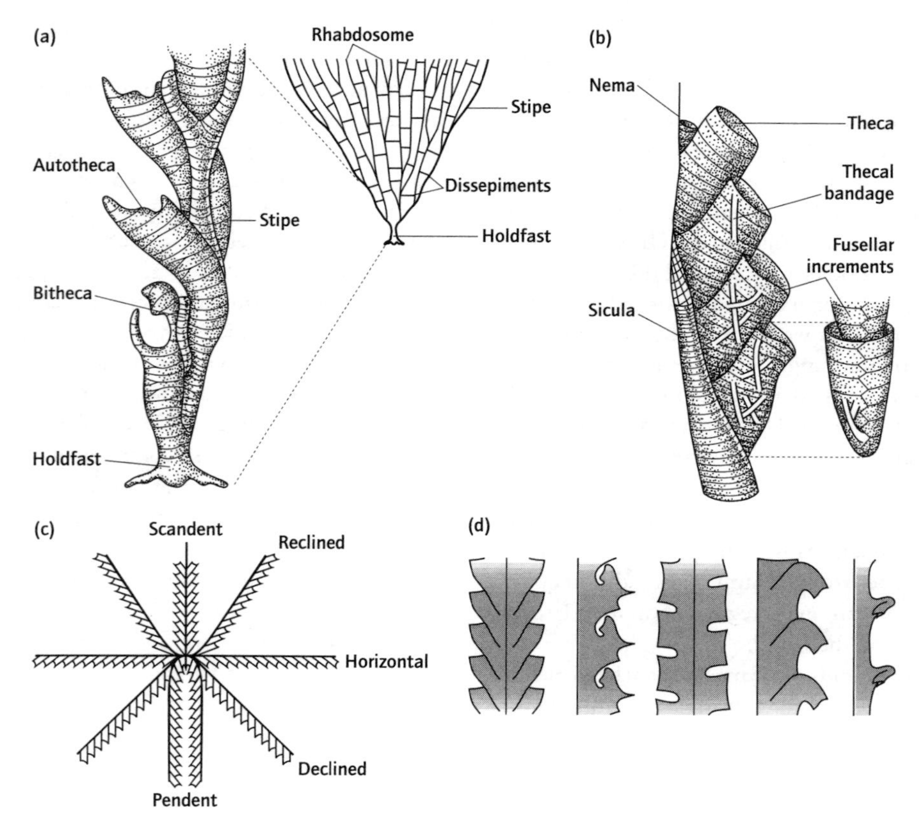

Fig. 10.1 The main elements of graptolite hard-part morphology: (a) dendroid; (b) graptoloid; (c) terms for describing the orientation of a graptolite stipe; and (d) different thecal types, left to right: glyptograptid, dicranograptid, climacograptid, hooked monograptid, enrolled.

Pterobranchs – the living relatives of graptolites

There are only two genera of pterobranchs, *Rhabdopleura* and *Cephalodiscus* (Fig. 10.2). They have a fossil record that extends back to the Middle Cambrian, and it was probably then that both they and graptolites evolved from a common ancestor.

Living pterobranchs are known worldwide, but are usually small and are easily overlooked. They are found from intertidal areas to abyssal depths, in water of normal salinity. Their preference is for areas with a rapid flow of water, and they often stick to boulders or dead shells in current-winnowed channels.

Both forms are colonial filter-feeders, but they are quite dissimilar to one another. *Cephalodiscus* grows a transparent, collagenous colony within which zooids move freely. One or several groups of zooids can be found in a single colony, with each group including between two and 20 zooids in various stages of development. The mature zooids are about 1 mm in size and are attached to the rest of the group by a contractile stalk. They have a simple body and an unusual head, from which sprouts a ring of filtering tentacles, called a lophophore, that forms a spherical filtering array. The head also includes the cephalic shield, a highly evolved and unusual organ used to secrete the skeleton and to hold the zooid in place.

Rhabdopleura grows transparent to brown colonies in which each zooid has its own "theca" and is connected to all the other zooids in the colony by a thin tube of tissue. Each zooid has a cephalic shield and a pair of filter-feeding arms. The colony form is variable, from single strands to complicated bushes. The most common type is a series of tubes that always grow in contact with another tube, forming a complicated two-dimensional maze on the surface of a shell or boulder.

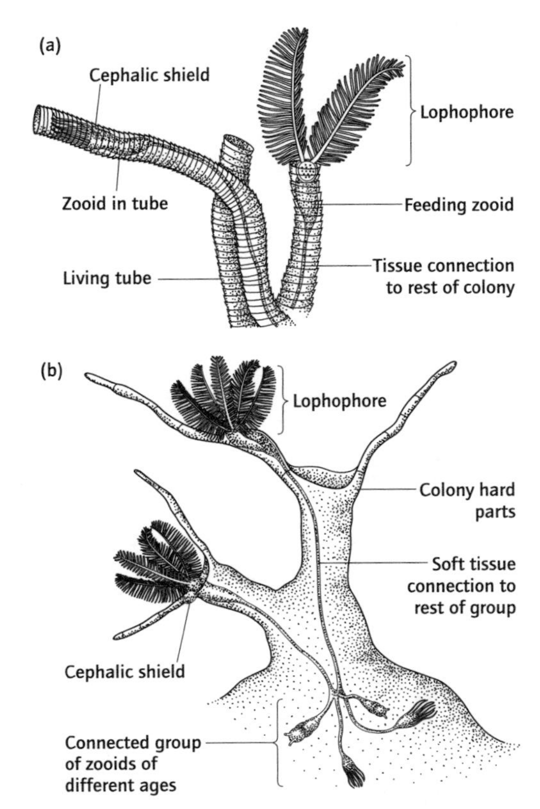

Fig. 10.2 Drawings of living pterobranchs: (a) *Rhabdopleura*, and (b) *Cephalodiscus*. The zooids are approximately 1 mm long.

Graptolite mode of life

Graptolite zooids are thought to have been similar in appearance to modern pterobranch zooids, and in particular to those of *Rhabdopleura*. A cephalic shield would have been needed to secrete the collagenous bandages seen on the surface of graptolite rhabdosomes, and the fact that this surface was accessible to the zooids suggests the absence of external soft tissue.

In planktonic graptoloids one of the most noticeable adaptations is to an extremely regular colony shape and it seems likely that this was an adaptation to living freely in water. In a planktonic mode of life the dominant control is often hydrodynamic – that is, the need to stay afloat and in the correct orientation with respect to food-bearing currents. Graptoloids achieved this in a variety of ways. Many independently evolved morphologies of graptoloids were designed to rotate as they moved, improving stability and increasing the amount of water sampled by each zooid. For very simple shapes (such as biserial forms and monograptids), stability, and the correct orientation with the sicula facing into a current, was achieved by having a nema. This acted like a tail on a kite, keeping its position downstream of the main colony.

The distribution of graptoloids seems to have been correlated with nutrient levels in the Lower Palaeozoic oceans. In many sections across the ocean shelves, graptolite diversity is highest near the shelf edge where upwelling currents would have supplied nitrates and phosphates to the primary producers on which graptoloids fed. In some areas, graptoloid blooms – similar to the blooms seen in modern plankton – can be identified, associated with thin volcanic ash bands. These bands represent eruptions that would have supplied vital micronutrients, especially iron and zinc, to the oceans.

Table 10.1 The main evolutionary events in graptolite history, along with illustrations of the dominant faunas and notes on important events of the time.

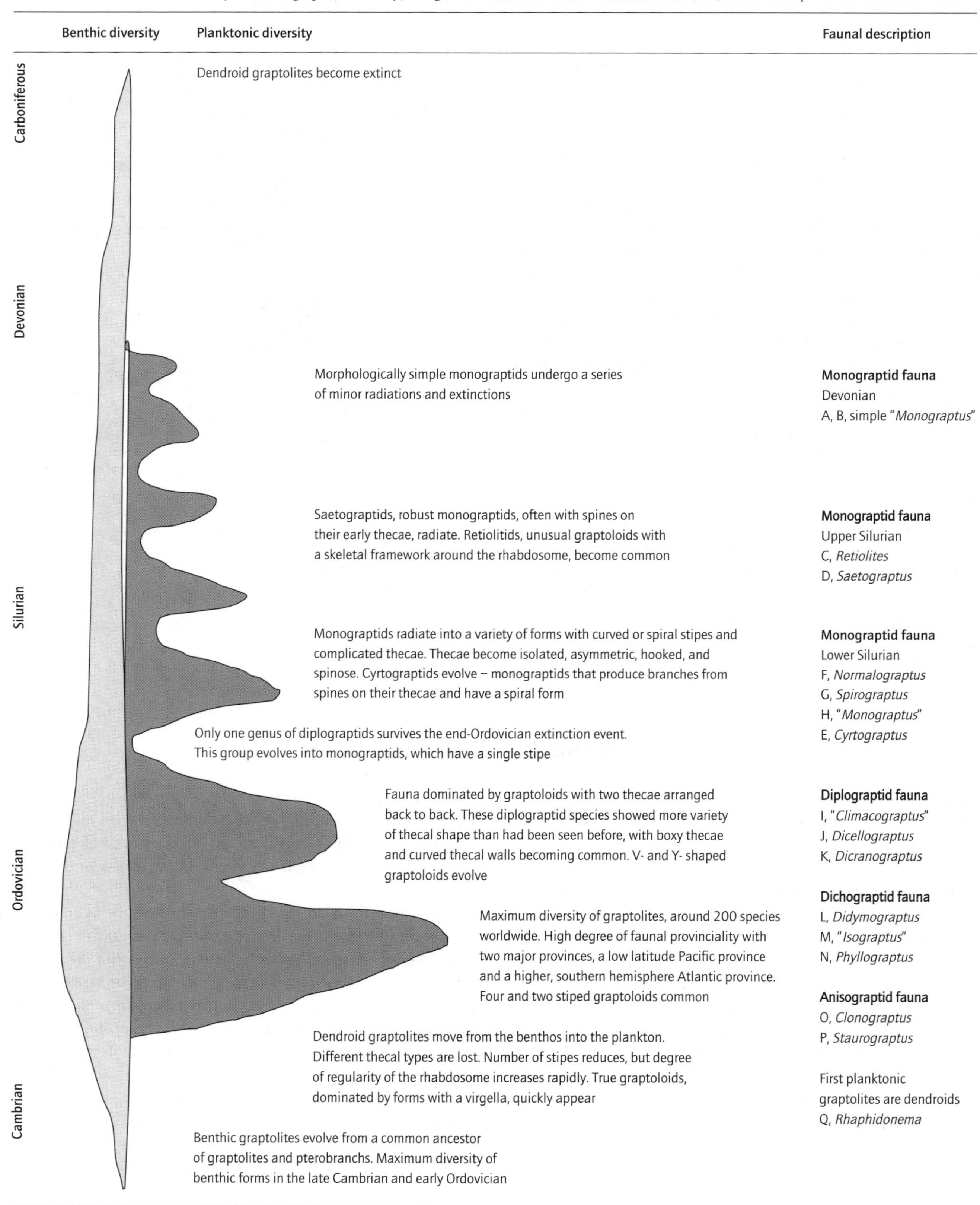

Representative graptolite species	A wider perspective

Dendroid graptolites gradually became less common and eventually became extinct, an event not associated with any mass extinction. Pterobranchs such as *Rhabdopleura* and *Cephalodiscus* are known from much more recent rocks, although their fossil record is extremely poor

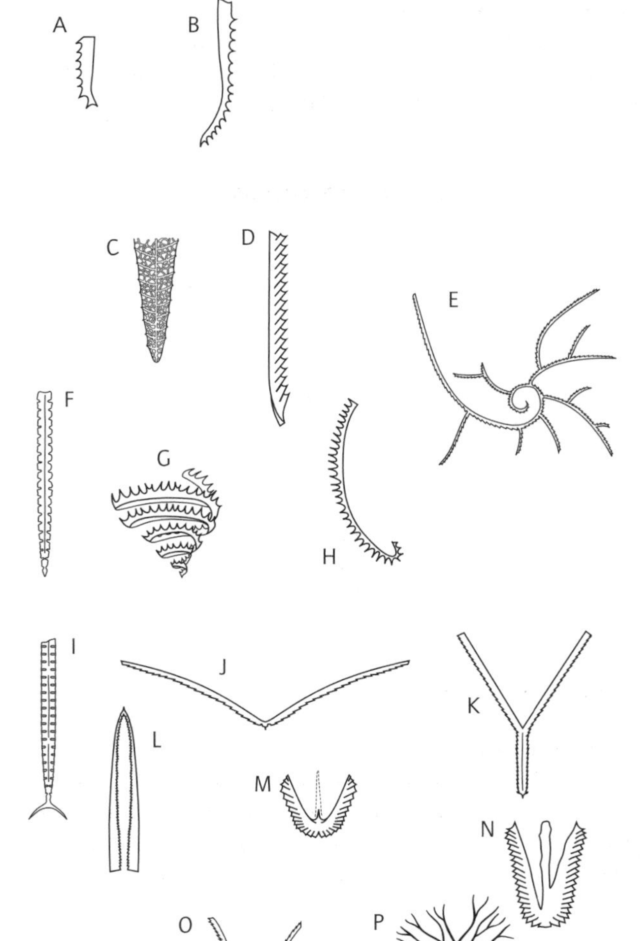

Major drop in diversity for all marine plankton. This was a mass extinction generated by a lack of species origination rather than an increase in rate of extinction. It is likely that this depression in diversity, which persisted into the Permian, was generated by the rise of land plants. These caused a radical reorganization of weathering patterns and sequestered enormous amounts of organic carbon and nutrients onto the land

Several smaller radiations and extinctions that happened in the Silurian may be related to climatic change. Deep water marine rock sequences of this age alternate between black shales, which indicate low oxygen conditions on the sea floor, and gray silts and muds, indicative of more oxygen. Oxygen is supplied to the sea bed via currents, so this alternation gives a measure of current activity in the seas. High diversity times in graptolite faunas correlate with periods of black shale deposition, when the oceans were poorly mixed. However, there may also be a degree of artifact about this observation, because graptolites are much more likely to have been preserved when scavengers were excluded from the sea bed because of the absence of oxygen

The end-Ordovician extinction was caused by a sudden and short-lived ice age. This event lasted perhaps 10 million years, and punctuated the longest warm period of the Phanerozoic. Large-scale extinctions occurred in low latitude plankton and benthos

Faunal provinciality was lost at this time, and never re-established, possibly limiting graptoloid diversity as most species have an extremely wide geographic distribution

Maximum graptolite diversity, of around 200 species, coincided with the existence of clearly developed faunal provinces. These were maintained by wide oceanic separation through the early Ordovician, when the Iapetus Ocean was at its maximum development

Dictyonema

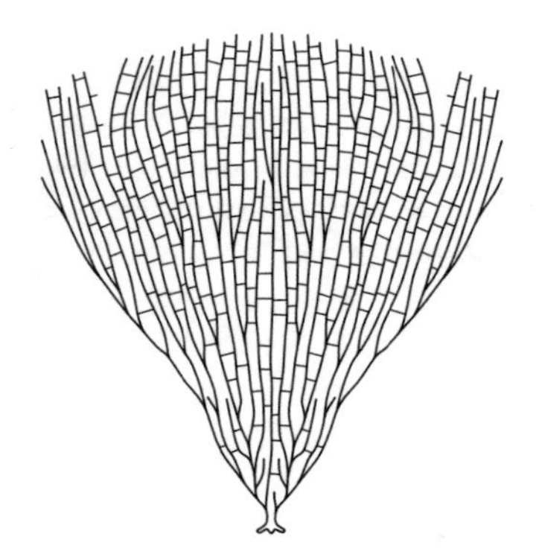

Cambrian–Carboniferous
A benthic dendroid graptolite up to 20 cm in height. Well over 1,000 zooids could have been accommodated in a fully grown colony. The rhabdosome is cone shaped, and held rigid by a series of dissepiments and attached to the sea floor by a holdfast.

Tetragraptus

Dichograptid fauna

Lower Ordovician
Four-stiped graptolite, in this instance with the stipes arranged like two U-shapes. They can be very large, with stipes up to 40 cm long, but are usually less than 10 cm in length.

Didymograptus

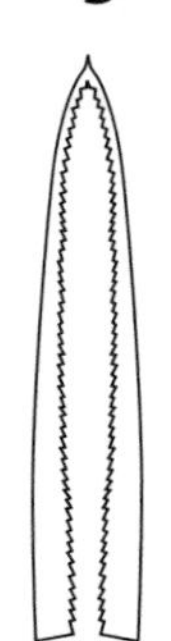

Dichograptid fauna

Lower–Middle Ordovician
Two-stiped graptolite, in this case with the two stipes pendent. They are generally 2–8 cm long. The thecae are simple and get larger for most of the length of the stipe.

Dicranograptus

Dichograptid fauna

Upper Ordovician
A Y-shaped graptoloid with distinctive thecae marked by two pronounced and abrupt bends, often with spines. They are typically 2–4 cm long, but can exceed 10 cm in length. The stipes have spirally arranged thecae after they separate.

Climacograptus

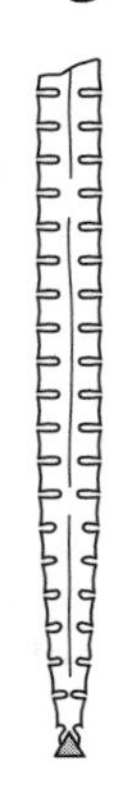

Diplograptid and early monograptid fauna

Ordovician–Silurian
A form genus, that is, a set of species that have a common shape rather than a common ancestry. These form genera are still useful for identifications in the field and at a preliminary level. Climacograptids have box-like thecae and are always biserial. They are typically 2–6 cm long.

Bohemograptus

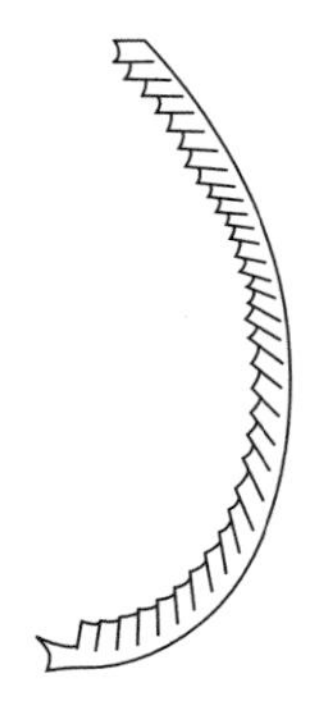

Monograptid fauna

Silurian
A single-stiped graptolite with simple thecae. The stipe can be curved or straight, usually in the order of 5–20 cm long and 2–3 mm wide.

Cyrtograptus

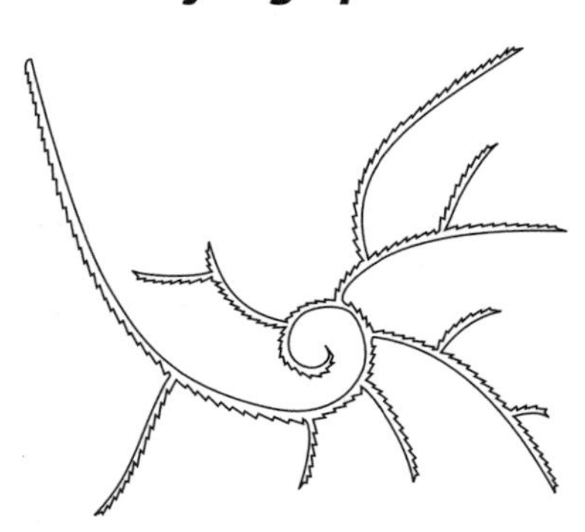

Monograptid fauna

Silurian
A spirally coiled graptoloid with unusual thecae, each with a spine at the aperture. These spines are capable of developing into branches, each made of the same type of thecae and capable of branching again. These species may have filled a niche left vacant by the extinction of large, many branched, horizontally arranged dichograptids. They are usually 5–60 cm across.

Rastrites

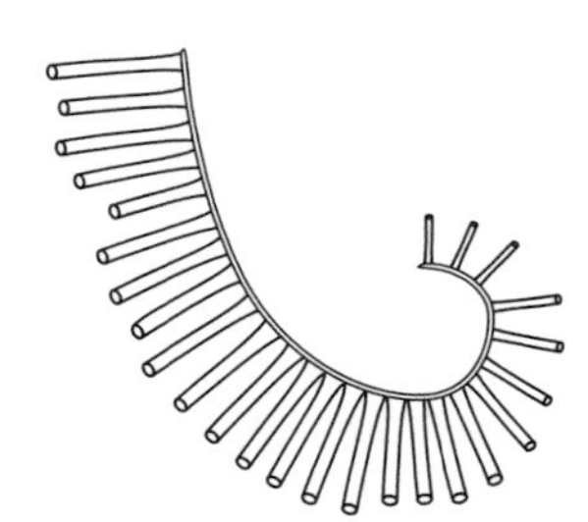

Monograptid fauna

Lower Silurian
Small, usually curved monograptids with extremely large, thin thecae each isolated from the next. The whole rhabdosome is usually less than 4 cm in length. The thecae can be up to 2 cm long and tend to have hooks or hoods at the apertures.

Glossary

Biserial – two-stiped graptolite, where the stipes are arranged back to back.
Cephalic shield – highly adapted organ possessed by living pterobranchs with which they secrete their skeleton and fasten themselves to it.
Collagen – fibrous protein characteristic of graptolites and pterobranchs.
Cortical bandages – strengthening elements of the graptoloid rhabdosome added to the inside and outside of the skeleton by zooids, presumeably using their cephalic shield.
Fusellar increments – unit of a theca added by a zooid to grow the colony. They are usually half rings with a zig-zag suture between.
Holdfast – modification to the apex of the sicula in benthic graptolites that enabled them to attach to the sea floor.
Lophophore – filtering array of tentacles possessed by pterobranchs, and assumed for graptolites.
Nema – thin spine extending from the top of the sicula.
Rhabdosome – name given to the entire graptolite colony.
Sicula – conical element of the colony, which is the first part to skeletonize.
Stipe – one branch of a graptolite rhabdosome, made up of numbers of thecae.
Theca – living tube for the graptolite zooid. Dendroids have several kinds, distinguished on the basis of size and function; graptoloids have one type.
Virgula – name sometimes given to a rod-like extension to the nema in scandent graptoloids.
Zooid – animal living in a graptolite rhabdosome or a pterobranch colony.

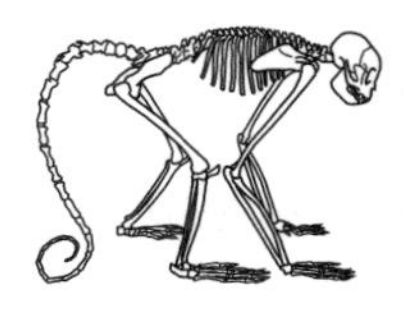

11 Vertebrates

- Vertebrates include the largest and fastest animal species.
- They were one of the few groups of organisms to colonize the land and the air.
- They are found in rocks throughout the Phanerozoic, from the Cambrian onwards.
- Vertebrates include fish, amphibians, reptiles, dinosaurs, mammals, and birds.
- The vertebrate skeleton appears uniquely adaptable to a wide range of functions.

Introduction

Vertebrates are of immense importance in aquatic and terrestrial ecosystems from the Ordovician onwards. They are commonly present as predators and scavengers, becoming almost ubiquitous in modern faunas. However, their characteristic, multipart, internal skeleton fossilizes poorly, and they are underrepresented in fossil assemblages.

The vertebrates are the most important group of the phylum Chordata. The characteristic element of a chordate is its notochord, the strengthened rod running down the back of the animal. This is usually mineralized in vertebrates to form a backbone, which surrounds a long nerve sheaf. In all skeletonized vertebrates the material used is calcium phosphate ($CaPO_4$), mixed with an organic material used as a template. This unusual and costly choice has physiological implications that may have contributed to vertebrate success. When organisms use oxygen faster than they can acquire it they begin to function anaerobically. This builds up high acidity in the body, which would quickly begin to corrode calcium carbonate secretions. However, calcium phosphate is resistant to such dissolution and allows vertebrates to overexert themselves for short periods, providing a useful energy boost.

The first vertebrates are Cambrian in age and include conodont teeth and rare fish. A close ancestor of vertebrates is the Burgess Shale animal, *Pikaia*. Successive radiations of different vertebrate groups have taken place, with fish becoming common and diverse in the Silurian, amphibians in the Devonian, different descendents of reptiles in the Permian, Triassic, and Jurassic, and birds and mammals in the Cenozoic (Fig. 11.1). Ecological niches have frequently been filled by successive groups of vertebrates, one after another, either competitively or by passive replacement.

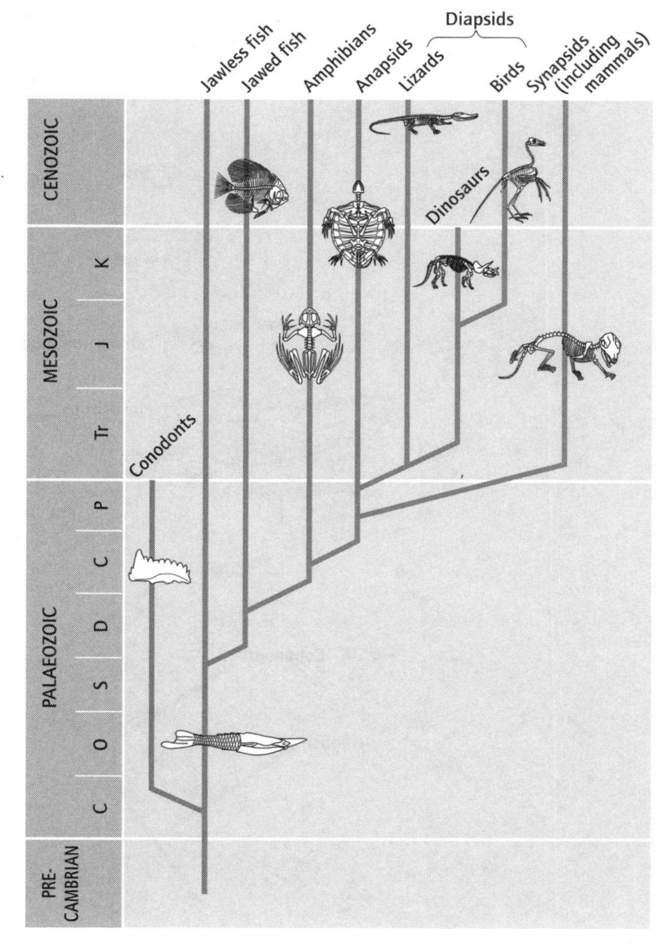

Fig. 11.1 The major evolutionary relationships of vertebrates.

Fish

The earliest vertebrates were fish, and all of them were marine, with fish not moving into fresh water until the Devonian. Most of these early fish lacked jaws. Any mineralization was concentrated on the teeth or on armor plating outside the body. Jaws evolved in the Silurian and this group, sometimes known as gnathostomes, quickly came to dominate fossil fish assemblages. Primitive gnathostomes evolved sequentially into the two most common modern fish groups, cartilaginous and bony fish, as well as into several extinct groups such as the placoderms and acanthodians. One group of bony fishes, the lobefins, evolved in turn into tetrapods and are our direct ancestors (Fig. 11.2).

The most commonly preserved early vertebrates are conodont animals, a hagfish-like primitive fish. These organisms are known best from their teeth, the microfossils called conodonts (for more details see Chapter 13). Conodont animals were all marine and their remains are found mainly in shallow water sediments from the Cambrian to the Triassic. Their teeth were made from dentine and enamel, but represent the only mineralized elements of the skeleton. Soft-bodied remains show that these organisms were active swimmers, with an elongated body between 5 and 50 cm in length, large eyes, and no jaws. They were almost certainly predators or scavengers, although reconstructions have been attempted showing the conodont elements arranged for filter-feeding. Wear patterns on well-preserved conodont elements are consistent with the pattern expected for grasping and cutting teeth.

Other primitive fish are known from Cambrian and Ordovician rocks, all of them lacking preserved jaws. Together, these jawless fish are known as agnathans, and they formed a common element of Lower Palaeozoic fish faunas. Today the group is represented by hagfish, which are scavengers, and lampreys, which live parasitically. Earlier species were probably more diverse in their ecology and may well have included some of the first marine predators. Others appear to have been well adapted to feed on sediment or on plankton.

The evolution of fish since the Silurian has led to significant changes in jaw apparatus and in the animal's ability to swim accurately and at speed. Paired fins, a strong, flexible body, and a mouth that can be stuck out to suck up food, are all adaptations that have appeared within the group.

Fishes with jaws, the gnathostomes, evolved in the late

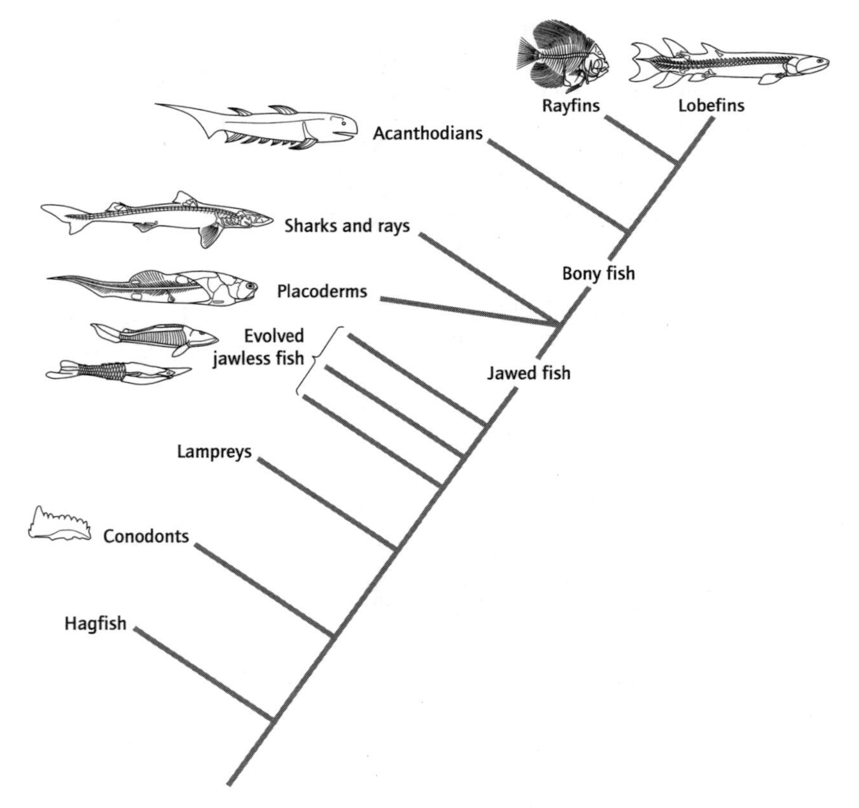

Fig. 11.2 The evolutionary relationships of fish.

Silurian. Jaws probably originated from the bony arches that support the gills, and may have been modified in order to improve the flow rate of water through the gills as much as for feeding purposes. However, this argument is based on modern fish that may be poor analogs for their Palaeozoic ancestors.

Gnathostomes diversified rapidly across the Silurian–Devonian boundary, making evolutionary relationships difficult to determine. Fish faunas of the time included heavily armored placoderms and the lightly armored acanthodians, characterized by having spines supporting the front of each fin. Placoderms such as *Dunkleosteus* reached 10 m in length and had a predatory habit. Some acanthodians had spikes on their gill arches, which might have allowed them to sieve plankton from water as modern baleen whales do. Primitive jawed fish were common in the Devonian, and are known from freshwater and marine sediments worldwide. They declined thereafter, and the last acanthodians became extinct in the end-Permian mass extinction.

Modern sharks and rays have a skeleton built from cartilage that is not mineralized. However, their teeth and scales are characteristically vertebrate as is their overall body plan. These fish are known as chondrichthyans and their first clear appearance in the fossil record is Devonian. Two radiations of sharks and their relatives have occurred, one in the Carboniferous and one in the Triassic/Jurassic. Most modern forms can be traced back to the Mesozoic, including predators and giant filter-feeders such as the basking shark.

Modern bony fishes, the osteichthyans, appeared in the late Silurian. During the Devonian they diverged rapidly into two main groups, the ray-finned fish that dominate modern aquatic environments, and the lobe-finned fishes. This latter group includes modern lungfish and the coelocanth, a famous "living fossil". Although rare in the modern world, this is the group that gave rise to all terrestrial vertebrates – including us.

Ray-finned fish have flexible fins supported by a lightly built fan of radiating bones, the rays. Most are rapid swimmers, and the power for this swimming is generally provided by movements of the body or of the tail, with the fins serving a steering function. Over time, these fish have tended to become more lightly built, and the external skeletal elements have been abandoned. There have been three main radiations of ray-finned fish, more properly known as actinopterygians. These took place in the late Palaeozoic, the late Triassic/Jurassic, and in the late Jurassic/Cretaceous. The latest of these radiations saw the spread of teleost fish, whose mouthparts can be pushed outwards to form a delicate sucking or plucking shape (Fig. 11.3). This innovation may have contributed to the success of the group, which now includes at least 20,000 species.

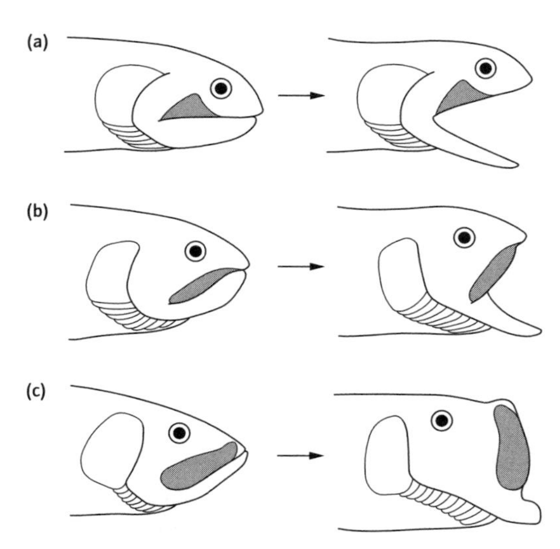

Fig. 11.3 Increasing flexibility of jaw action in the jawed fish: (a) an early jawed fish; (b) a primitive bony fish; and (c) an advanced bony fish.

Lobe-finned fish are known as sarcopterygians. Their fins are sturdy and supported by a few large central bones, usually supported by a strong linkage to the skeleton. In contrast to actinopterygians, the fins produce a power stroke to move the fish along. This strongly built, powerful fin is a useful pre-adaptation to life on land. Another useful adaptation is the ability to breath air. This is relatively common in fish, especially those living in warm, shallow water, which is prone to become stagnant. In this environment many fish will suck in bubbles of air from which they extract oxygen. Modern lungfish can breath air indefinitely. Sarcopterygians reached their maximum diversity during the Devonian, and also gave rise to amphibians in that period. They have formed a very minor component of fish faunas since that time.

Amphibians

Amphibians are tetrapods (four-limbed vertebrates) that lay eggs in water. They are the ancestral group to all of the other tetrapods, including reptiles, dinosaurs, and mammals, as well as birds.

The most likely ancestors of amphibians, and all other tetrapods, are a group of extinct lobefin fish known as rhipidistians (Fig. 11.5). These fish have a similiar skull morphology to the earliest amphibians and the pattern of limb bones common to all subsequent tetrapods, including humans (Fig. 11.4). This is the pattern of one upper bone, two lower bones, and many peripheral bones in each limb. For most tetrapods there are five peripheral bones, for example our fingers, but some species have modified this number and the earliest amphibians tended to have more than this, typically seven or eight digits.

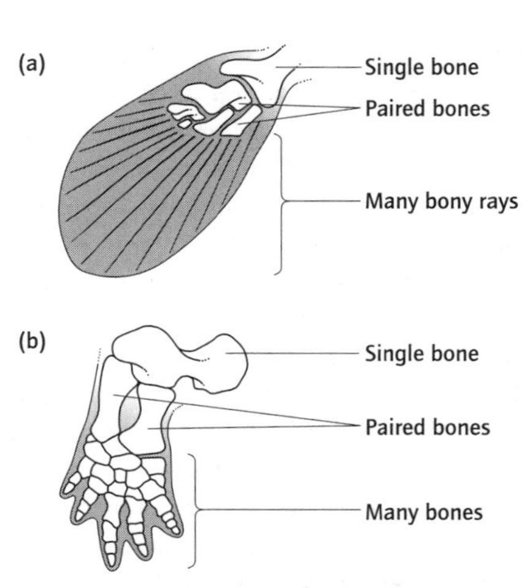

Fig. 11.4 A comparison of the fin of (a) a rhipidistian, and (b) the limb of a primitive tetrapod.

With their strong fins and ability to breath air, lobefin fishes were excellently preadapted to a life lived partly on land. In addition, they were adapted to shallow water habitats, in both the seas and fresh water. This put them physically close to land, in an environment that could intermittantly become problematic, during a dry season or when stagnation caused a lack of oxygen supply. Modern lungfish leave water to escape from such conditions and to find food. The ability to crawl over land might also have allowed lobefins to lay their eggs in isolated bodies of water for protection from scavengers.

The two best-known early amphibians are called *Ichthyostega* and *Acanthostega*. Both are found in late Devonian sediments from Greenland. They both appear to have been lake-living fish-eaters and were highly adapted for this way of life. However, they also show a fascinating blend of fish-like and amphibian-like characteristics that give a real insight into this stage of tetrapod evolution.

Both of the Greenland amphibians had streamlined bodies and heads, a long flexible tail, and a pronouned tail-fin. Their teeth were very similar to the teeth of fish. Their limbs were short and the wrists and ankles poorly articulated. Oddly, *Icthyostega* had seven toes on its hindlimbs, and *Acanthostega* had eight digits on its forelimbs. This is important because it shows that the five-digit hand and foot, the so-called pentadactyl limb, is a later development and is not one of the shared characteristics of the whole group.

Amphibians radiated into around 40 families during the Carboniferous, thriving in the warm, damp coal forests. The diversity of form and size make these organisms hard to classify. They include snake-like animals, tiny amphibians with adults less than 10 cm long, and aquatic forms with wide, delta-wing heads. From this highly diverse assemblage two groups are important for the evolution of more modern tetrapods. These are the temnospondyls, that gave rise to modern amphibians, and the reptilomorphs, that probably include the ancestor of reptiles, mammals, and birds.

There were three things that limited the evolutionary potential of these Carboniferous amphibians: they needed to return to water in order to lay eggs, they were all meat-eaters, and they had limited mobility on land. Limbs were generally short and articulated at the knee and elbow, producing a low-slung, ungainly motion.

Modern amphibians are divided into three groups: frogs and toads, newts and salamanders, and caecilians, which lack limbs. They are classified together as lissamphibia, and are thought to have evolved from a temnospondyl ancestor during the Triassic. The earliest frogs are early Jurassic in age and have the characteristic shape of modern frogs. A salamander has been found in late Jurassic rocks from Kazakhstan. The oldest known caecilian is also Jurassic in age and still has small limbs. The fossil record of the lissamphibia is relatively poor.

Carboniferous reptilomorphs included groups such as the anthracosaurs, medium-sized fish-eaters around 1 m in length. Some anthracosaurs, such as *Protogyrinus*, had relatively long limbs and would have been fully mobile on land. Others, such as *Pholiderpeton*, were adapted for living in water. An unknown member of this group evolved the ability to lay eggs on land. These amniotic eggs were critical to the success of descendent groups, such as reptiles and mammals.

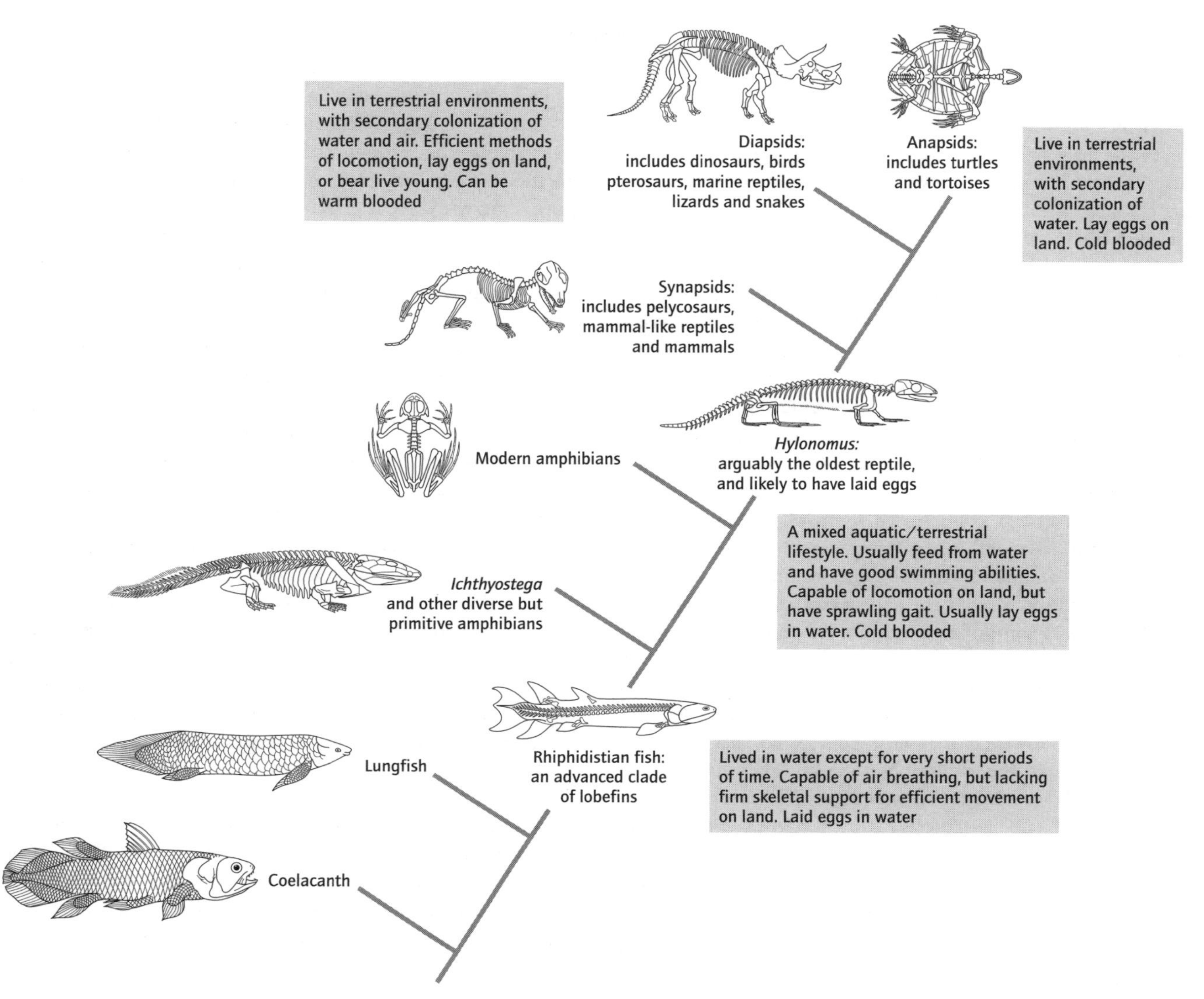

Fig. 11.5 The evolutionary relationships of tetrapods. The boxes show the main functional adaptations of each group.

Amniotes

Reptiles evolved from amphibians during the Carboniferous. Their key innovation is the ability to lay eggs on land. These amniotic eggs are a life support system for the embryo away from water. They have allowed reptiles and their descendants to colonize a wide range of environments not accessible to amphibians.

An amniotic egg develops from a single fertilized cell. It contains the yolk, which develops into an embryo, and its food supply (the "white"). In addition there is a space for storing waste products and an air pocket through which gases exchange with the air outside. The whole package is wrapped in a waterproof membrane that can be rigid or flexible (Fig. 11.6).

Unfortunately, eggs are very rare in the fossil record and the oldest known egg is Triassic in age. However, all eggs produced by modern amniotes are generated in the same way. It is considered that this complicated series of processes is most likely to have arisen only once in evolution, by definition in the most primitive amniote. Other details of amniote anatomy can therefore be used to infer the appearance of egg-layers in the fossil record. A suite of early reptiles is known, any of which could have been ancestral to the group. One possible early amniote is *Westlothiana*, a small tetrapod found in a Carboniferous volcanic lake deposit from the Midland Valley of Scotland. However, it is now considered that this form slightly predates the evolution of eggs. The earliest well-known reptile is called *Hylonomus* and is found in the hollow tree stumps of a Carboniferous fossil forest in eastern Canada.

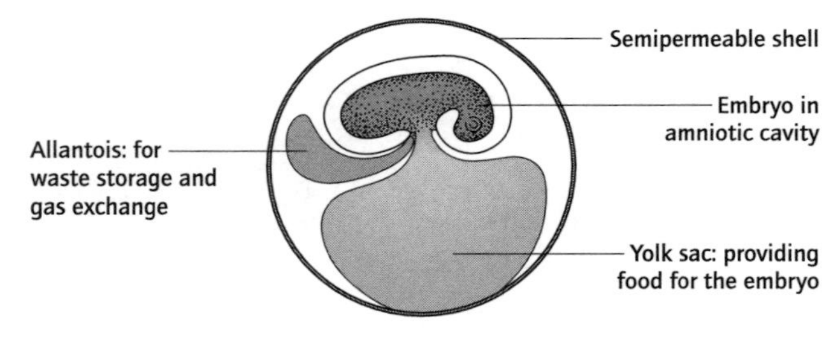

Fig. 11.6 The amniotic egg, which has allowed advanced tetrapods to fully colonize the land.

Early reptiles were small in size, and this may be a reflection of their physiology. Modern reptiles are cold blooded, and need to maintain a functional body temperature by behavioral methods. They may bask in the sun, or need to burrow to escape from excessive heat, for example. Such temperature regulation is much faster for a small animal, with a high surface to volume ratio. In contrast, the early amphibians could be much larger, as they lived mainly in water where temperature control is easier.

During the Carboniferous, reptiles radiated into the three main evolutionary lines, or clades, that have dominated terrestrial environments since. This early radiation was into distinct lineages, which are most characteristically differentiated by their skull structure. The anapsid lineage is primitive and has no holes in the upper skull apart from the eye and nose apertures. This lineage is represented today by modern turtles and tortoises. It has occasionally been more diverse than at present, but has never rivaled in importance the two lines that evolved from it. The synapsid lineage evolved next, and had a single hole in the upper skull behind the eye. This group evolved into mammal-like reptiles and eventually into mammals. The diapsids evolved later and are characterized by two openings in the skull behind the eye. This group diversified into most modern reptile groups, marine reptiles, dinosaurs, and birds (Table 11.1).

Each advanced family of reptiles developed different solutions to the problem of greater mobility on land. Amphibians have limbs that articulate at the knee and elbow. Their pelvic and shoulder girdles are rigid and the backbone between flexes to one side and then the other as they walk. This limits their speed and also the length of time for which they can move. As the spine flexes, it compresses one lung and then the other, making it impossible to breath and walk at the same time. This limitation also applies to modern reptiles, such as crocodiles, that can lunge at prey, but not chase a prey animal over any distance. Synapsids and diapsids have at different times solved this problem, known as Carrier's constraint, in a variety of different ways.

Table 11.1 The three main clades of reptiles.

	Origin	Main radiations	Common examples
Anapsids No skull opening behind the eye socket	Carboniferous origin; this is the primitive group of reptiles	Never very abundant or diverse Greatest diversity of form in the Permian Greatest success after the evolution of the shell in the Triassic	Turtles, tortoises
Synapsids One skull opening behind the eye socket	Carboniferous	Pelycosaurs in the early Permian, therapsids in the late Permain True mammals in the Palaeocene	*Dimetrodon*, kangaroos, horses, humans
Diapsids Two skull openings behind the eye socket	Carboniferous	Archosaurs, marine and flying reptiles in the Triassic Dinosaurs in the Jurassic Birds in the Palaeocene	Ichthyosaurs, plesiosaurs, pterodactyls, *Triceratops*, *Tyrannosaurus*, seagulls

Anapsids

Anapsids are a group that includes the earliest reptiles, a smattering of Permian and Triassic forms, and the modern turtles and tortoises, whose fossil record also extends back to the Triassic.

Permian and Triassic anapsids fall into three families. The Permian millerettids are known from South Africa, which was at temperate southern latitudes during that time. They were small, active insectivores of moderate size, with skulls typically around 5 cm long. Late Permian and Triassic procolophonids lived in moderate to high southern latitudes and were omnivores or herbivores. Late Permian pareiasaurs are found in the northern hemisphere and could reach 2–3 m in length. They were heavily build herbivores.

This variety of feeding strategy is also seen in the modern anapsids, tortoises, and turtles, even though the group lacks teeth. All share the common feature of a protective shell into which the body can be withdrawn. It is built from two elements, the carapace on top and the plastron below. Bony plates grow within the skin on the back of the turtle, or tortoise, which are then covered with a horny plate or scute. The plates of the carapace attach to the ribs and backbone, while the plastron attaches to the shoulder girdle.

The oldest fossil turtles are late Triassic in age and they appear to have evolved either from the procolophonids or pareiasaurs. At their maximum diversity they reached 25 families. Fossils have been found from marine, freshwater, and terrestrial environments, and can reach great size, with carapaces growing to over 2 m in diameter.

Synapsids

Synapsid, mammal-like reptiles, dominated the late Carboniferous and Permian land masses. The first radiation was of the group known as pelycosaurs (Fig. 11.7). This group moved into a range of dry habitats, and evolved to large size, and in some cases to a herbivorous diet. The best known pelycosaur is *Dimetrodon*, with a skeleton characterized by a huge sail supported by extensions to its backbone. This sail probably helped the animal to maintain its preferred body temperature, and was one of a range of innovations that contributed to the increase in size and ecological diversity of the group. Descendents of the pelycosaurs, therapsids, radiated widely in the late Permian and extended the domain of tetrapods into much higher latitudes. These organisms were shorter and more squat than the pelycosaurs, and were probably able to generate some of their body heat internally, that is they were probably warm blooded to some degree. Their descendents, the cynodonts, were common in the Triassic, and include the species *Thrinaxodon*, which shows evidence of having had whiskers. As whiskers are modified hairs, it is probable that the group was furry, which is only of benefit to warm-blooded organisms.

Although mammal-like reptiles were advanced in many ways, they all shared a sprawling gait with their amphibian ancestors. This would have limited their success when competing with a more mobile organism, as many evolving diapsids were. Synapsids were outcompeted by the ancestors of dinosaurs during the Triassic, and remained a minor group until after the end-Cretaceous extinction event.

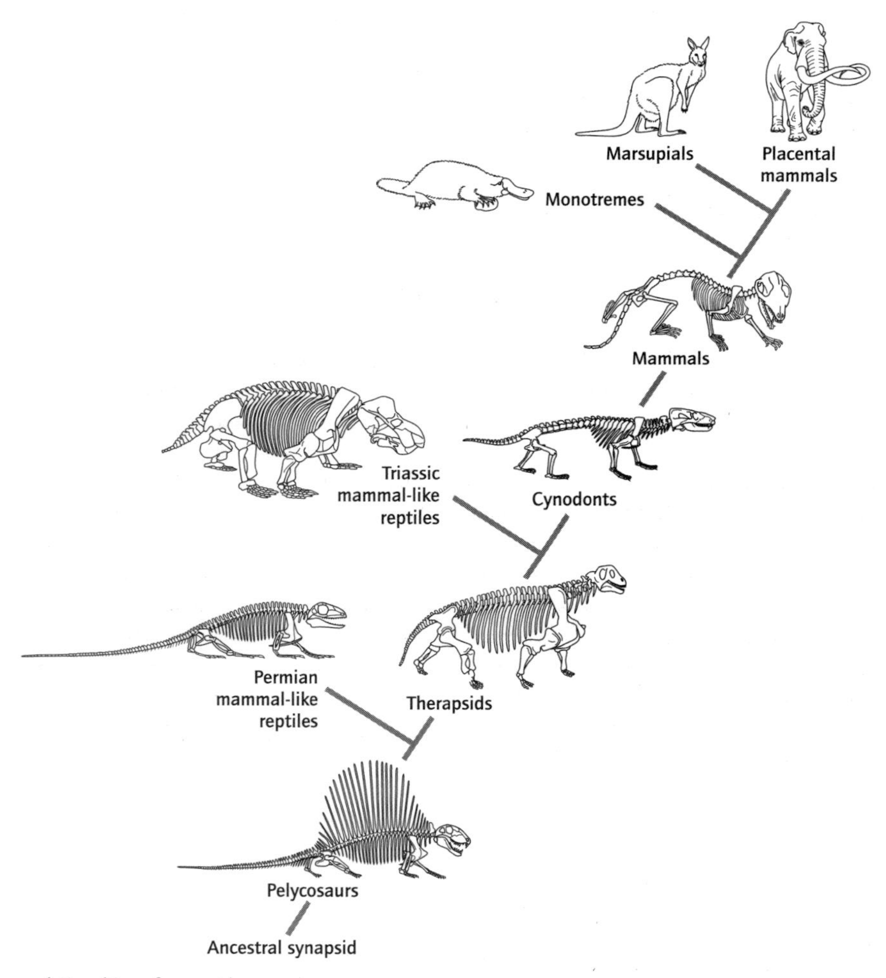

Fig. 11.7 The evolutionary relationships of synapsids.

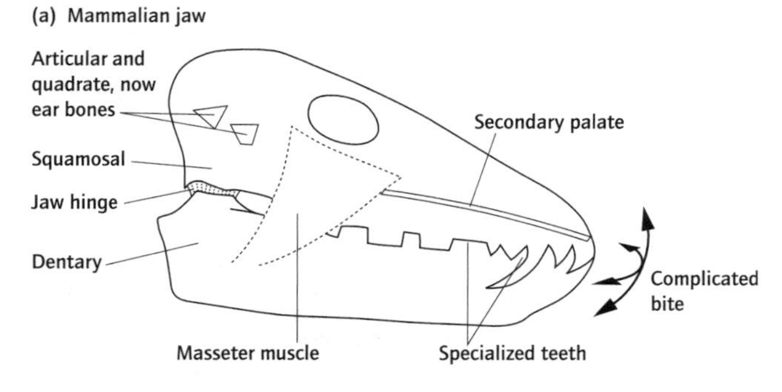

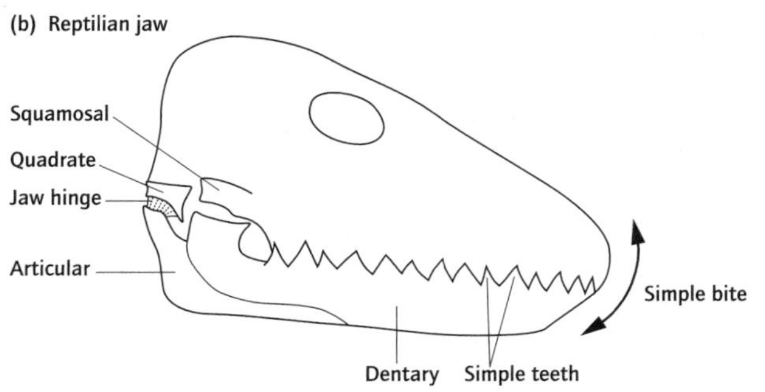

Fig. 11.8 Simplified diagram indicating the main differences in jaw structure between (a) mammals, and (b) their ancestors. These characteristics were acquired gradually and intermediate species can be identified displaying a "mosaic" of primitive and advanced characters.

Mammals

A cynodont similar to *Thrinaxodon* evolved into mammals during the Jurassic. The acquisition of mammalian characters was a patchy affair, and it is difficult to identify a meaningful point where the ancestral group could be said to have evolved into its descendent. The main changes involved are in mobility and jaw structure.

True mammals were the first synapsids to evolve an erect gait, that is, to articulate their limbs at the shoulder and hips. By doing this, they were able to overcome Carrier's constraint, and to move and breath at the same time. Mammals were also aided in overcoming this problem by the evolution of a diaphragm to protect the lungs from compression.

The jaw array of true mammals has highly specialized teeth and a large cheek, with the chewing action controlled by a major muscle, the masseter, running between the jaws. The lower jaw is composed of one bone, the dentary, that articulates with an upper squamosal bone. In mammal-like reptiles and in diapsids, two extra bones are involved in jaw articulation, the articular and quadrate bones. In mammals these have migrated into the ear where they allow airborne sound waves to be picked up and transmitted to the brain (Fig. 11.8).

The most familiar feature of modern mammals is the ability to bear live young. However, monotremes such as the platypus lay eggs, and marsupial and placental mammals have very different styles of gestation. Many nonvertebrate groups have independently evolved the ability to give birth to live young. This feature thus assumes less importance in the identification of the clade, and possibly in explaining its Cenozoic success.

Most Mesozoic mammals are known only from their teeth, which have local stratigraphic importance, or from incomplete skulls. They were all small in size and may well have been nocturnal. The majority were predators or omnivores. They included the three subclasses seen today – the monotremes, marsupials, and placental mammals – along with a variety of extinct groups. Most groups were confined to a restricted geographic region. This restriction was enhanced by the breakup of the supercontinent Pangea during the early evolution of mammals. Thus monotremes, which lay eggs, are found only in Australasia; marsupials, which suckle their young in pouches, are found in South America and Australasia; and placental mammals, which retain their young for longer inside the body, are found in Asia, Europe, and North America.

Sixty-five million years ago, dinosaurs became extinct during the end-Cretaceous mass extinction. Within 10 million years, mammals had radiated into almost all of the familiar modern families, including primates and whales. A great deal of parallel evolution occurred, with similar placental and marsupial mammals evolving independently on different continents. A sabre-toothed marsupial is known from South America, for example, very like the famous placental sabre-tooths of the last ice age.

The first large land animals to colonize the post-dinosaur world were a bizarre set of organisms. They included flightless birds, much larger than modern ostriches, which were predators as well as grazers. Enormous crocodiles, 10 m in length, also competed for the top predator niches. The mammals that radiated alongside these forms were predominantly small carnivores, scavengers, and herbivores. The larger mammalian carnivores that gradually appeared during the Palaeocene and Eocene belonged to an extinct group of mammals known as the creodontids. For the first 10–15 million years of the Cenozoic, faunal interchange on land was extremely limited, leading to the independent evolution of a wide range of mammals in each continent. These radiations included most of the modern groups of mammals, such as modern carnivores, whales, bats, and hooved grazers.

A major period of climatic cooling began in the Eocene. This had a dramatic effect on vegetation patterns, which caused stress in many terrestrial ecosystems. Falling sea levels, accompanying the development of ice caps at the poles, produced land bridges connecting previously isolated land masses. The net effect of these changes was to cause wide-spread extinction amongst mammal families, with the survivors being predominantly the modern forms with which we are familiar.

Primates and humans

The order Primates, to which we belong, also includes lemurs, monkeys, and apes, as well as our direct ancestors. The group can be traced back to the late Cretaceous, but radiated in the well forested, low latitude environments of the early Cenozoic. The ancestors of apes and humans are thought to have evolved about 25 million years ago, at a time when the planet was drying and cooling as ice began to build up at the poles. This trend was even more marked in Africa because of local climatic effects caused by its northward drift and the closure of the Tethyan Ocean to the north.

Around 6 million years ago, as grassland replaced forests, the ancestors of humans moved out onto the plains, and in doing so acquired an upright stance, freeing their upper limbs for carrying food or tools. The earliest of these upright, grassland-dwelling apes were the australopithecines. Footprints have been found in ash falls dated at around 4 million years old, and the oldest well-preserved skeletons, of *Australopithecus afarensis*, are dated at 3.2 million years ago. Australopithecines evolved in two different ways, towards heavily built vegetarians with small brains, and towards more lightly built omnivores who used simple tools and communal behavior for survival. The more robust lineage is usually known as *Paranthropus*. Various species of *Paranthropus* have been identified in Africa, and it is likely that groups of these large vegetarians coexisted with our direct ancestors. The less robust strand of *Australopithecus* evolved into the genus *Homo*. This genus evolved through the species *Homo habilis* and *Homo erectus* into *Homo sapiens*, a species that includes Neanderthal man as well as ourselves (Fig. 11.9).

As physical changes occurred in the hominid skeleton, a comparable development in behavior is seen, preserved as tools and in the fossil skeletons themselves. Tool-making, as distinct from using found objects as tools, is unique to humans, and developed into a progressively more sophisticated form through hominid evolution. Greater ranges of material were employed to construct the tool kit, and the range of functions that tools served increased as well. Analysis of brain shape and the structure of the throat suggests that earlier species of hominid may have been capable of speech, though it may have been more limited in diversity than that acquired by modern humans. The burial of the dead, accompanied by ceremony, is known from Neanderthal sites, and art in a variety of types is known from the sites occupied by modern humans from about 40,000 years ago.

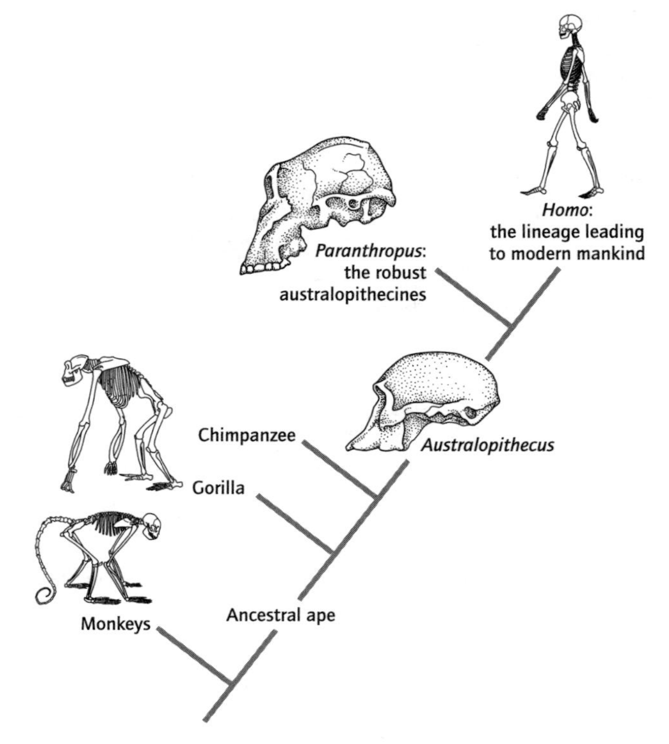

Fig. 11.9 The evolutionary relationships of primates.

There is great debate about the exact nature of hominid evolution, and much of it is based on sparse evidence. In particular, the timing of hominid migration out of Africa has caused much dispute, as has the point at which the last common ancestor of all humans lived. Table 11.2 summarizes the basic information about the most important species of hominid, tracing characteristic changes in brain size, height, and behavioral complexity.

There is no great "missing link" between apes and humans, and the fossil record, though poor, confirms that our evolution has been similar in all respects to the evolution of trilobites, molluscs, or fish. The cooling of Africa and subsequent spread of ice across the northern hemisphere was not simply a background for human evolution, but its crucible.

Table 11.2 The main features of well-known species of hominid. The table shows the physical features of these species, their distribution in a changing physical world, and evidence of their mental development as shown by preserved artifacts or skeletal structure.

	Physical features	Location and climate	Development
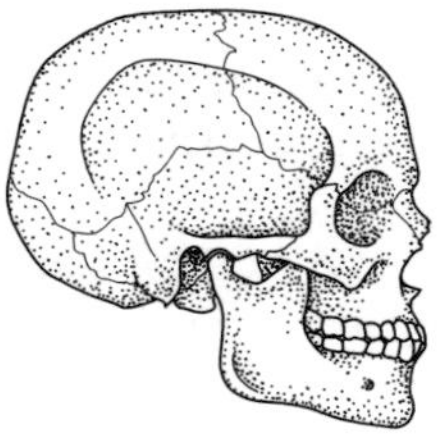 *Homo sapiens sapiens*	Brain size: 1,500 cm^3 Height: 1.7–1.8 m Anatomically modern humans may be found from 90,000 years ago and are common after 40,000 years ago	Known from sites worldwide, including America and Australia. May have lived in predominantly warmer sites well south of the major ice sheets	Stone-age communities produced diverse tools and artifacts, including painted and carved images. Active hunter gatherers
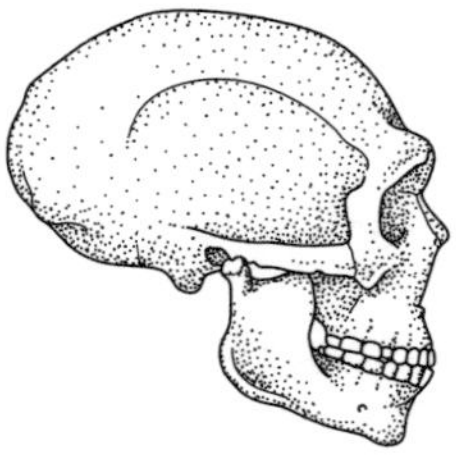 *Homo sapiens neanderthalensis*	0.2–0.035 Ma Brain size: 1,500 cm^3 Height: 1.6–1.7 m Thick set humans with strong limb bones and a brow ridge	Adapted to cold conditions. Best known from around the margins of the ice. Widely found in Europe and the Near East	Mousterian tools, with a great diversity of elements. Use of fire and caves. Evidence of burial of the dead. Actively hunted large mammals
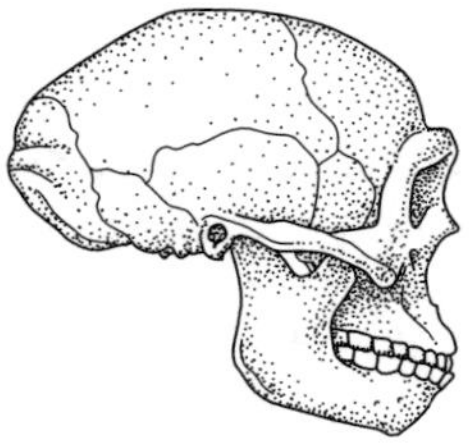 *Homo erectus*	1.7–0.5 Ma Brain size: 1,100 cm^3 Height: 1.3–1.8 m Physically like modern humans except for differences in skull architecture, especially a brow ridge	Known from Africa, Asia, and Europe. Probably nomadic. Lived during a period of climatic deterioration with many land bridges produced as ice caps formed	Produced Acheulian tools, which included axe heads and spears. A wide diversity of materials were used. Fire is known from campsites; cave dwelling is known
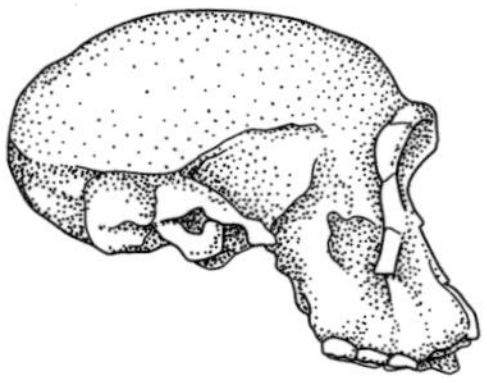 *Homo habilis*	2–1.5 Ma Brain size: 700 cm^3 Height: 1.0–1.5 m Limbs and body proportions similar to modern humans. May have had speech.	Only known from African savannah settings	A toolmaker of the Oldowan set of tools – mainly flakes from chert or flint. These enabled *H. habilis* to actively hunt and scavenge, producing an omnivorous diet
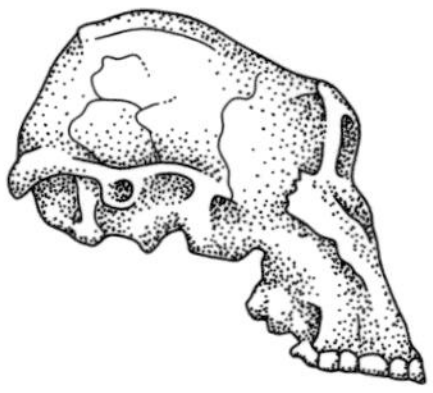 *Paranthropus boisei*	2–1.2 Ma Brain size: 550 cm^3 Height: 1.1–1.5 m Massive teeth designed for rough vegetation, and a robust body form. Fully adapted to life on the ground	Only known from Africa, living mainly on grassland terrains. Climates worldwide were cooling and drying at this time, with marked changes in vegetation	Had a good grip, and probably used found objects as tools. A vegetarian diet required extensive time to collect, possibly limiting social development
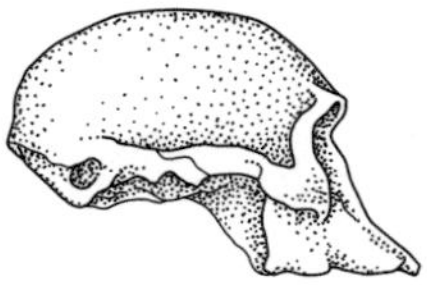 *Australopithecus afarensis*	3.0–2.5 Ma Brain size: 350 cm^3 Height: 1.1–1.5 m Many ape-like characteristics, but able to walk upright. Probably lived on the ground and in trees	Only known from Africa, at a time of widespread open woodlands and equable climates. Probably lived between the woods and the plains	May have used found objects as tools. Vegetarian diet, especially relying on fruit and leaves from trees rather than the savannah grassland

Diapsids

In order to follow the evolution of the other major group of reptiles, the diapsids, we must return to the Triassic. Early diapsids had evolved into small or medium-sized carnivores, but had not been able to compete successfully with mammal-like reptiles. However, following the end-Permian extinction event, they radiated significantly and during the Triassic replaced synapsids by active competition in almost all ecological niches. This successful group of diapsids is called archosaurs. It may be that they were able to outcompete the mammal-like reptiles because they developed a range of solutions to Carrier's constraint. Many evolved a bipedal stance, and from that an erect gait that permitted them to move and breathe at the same time. Evidence suggests that primitive archosaurs may have been warm blooded. Endothermy requires a lot of energy, and it reduces the number of predators able to live on a given number of prey animals. The predator : prey ratios of Triassic ecosystems are intermediate between modern warm- and cold-blooded predators.

Most archosaurs became extinct in the late Triassic, and their ecological niches were filled by their descendents, the dinosaurs. At the same time the first flying vertebrates appeared, the pterosaurs. Dinosaurs evolved from small, carnivorous archosaurs and by the end of the Triassic they had radiated into three main groups: the theropods, which are primitive in some ways, the sauropods, and the bird-hipped dinosaurs or ornithischians (Fig. 11.10).

The Mesozoic is often called the "Age of the Dinosaurs", but it might better, though less catchily, be called the "Age of the Diapsids". Marine reptiles in the ocean, pterosaurs in the air, and dinosaurs on the land dominated the middle and upper parts of all ecosystems, representing the top predators and often the lesser predators, omnivores, and herbivores as well. More has been written and is popularly known about these groups than about any other fossils. It is easy to forget their general rarity, and the major uncertainties that still exist about their life habits and appearance.

Pterosaurs ranged in size from a wingspan of a few centimeters to over 15 m. They were active flyers, and had narrow, membranous wings attached to a modified fourth finger and probably to their thighs. The aerodynamics of their wings suggest that they were predominantly gliders and soarers, but rarer species were adapted to other flying habits. They were covered in hair, and their highly energetic lifestyle means that they must have been warm blooded. Most pterosaurs appear to have been coastal or marine predators, although this may well be an artifact of preservation, with these being the most

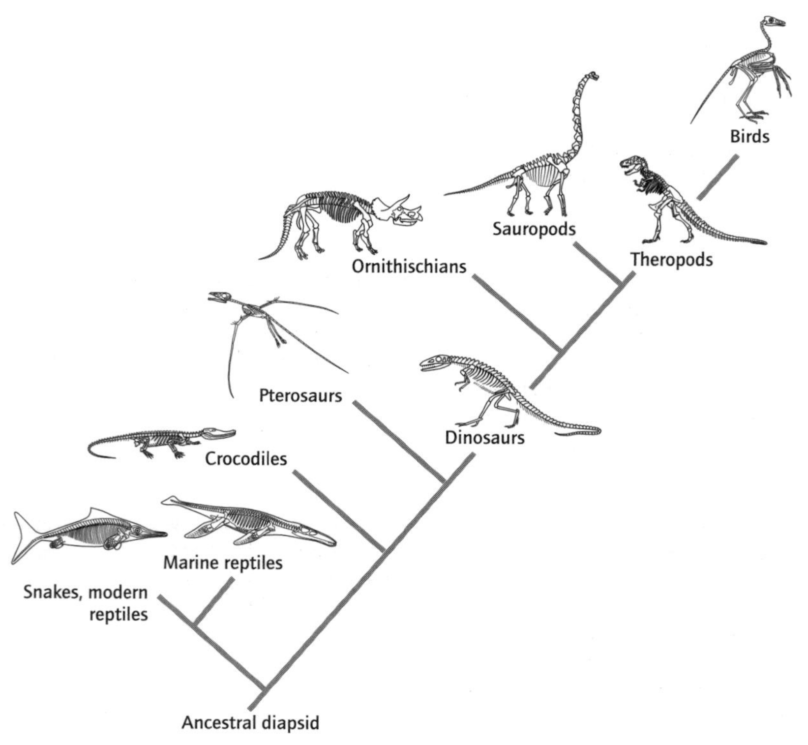

Fig. 11.10 The evolutionary relationships of diapsids.

likely environments in which to preserve a skeleton. In common with dinosaurs, they became extinct at the end of the Cretaceous, though the physiologically similar birds survived.

Marine reptiles appeared in the Triassic, and radiated through much of the Mesozoic. Some groups that took to the water are familiar today, such as turtles and crocodiles. Other are extinct, and the most important of these are the ichthyosaurs and sauropterygians, including the plesiosaurs and pliosaurs. The origin of ichthyosaurs is unknown; they are a highly evolved, or derived, family. They had large eyes and sharp, conical teeth, and well-preserved specimens often contain fish scales or belemnite hooks in their stomachs. Rare specimens have been preserved in the act of giving birth, showing that they bore live young. Sauropterygians evolved from the line leading to modern reptiles and snakes in the Permian. They evolved into marine predators that swam using their highly modified limb paddles. Some may have been specialized bottom-feeders while others were predatory on smaller marine reptiles. Along with pterosaurs and dinosaurs, most marine reptiles became extinct at the end of the Cretaceous.

Dinosaurs

The most famous Mesozoic diapsids are the dinosaurs. Theropod dinosaurs appeared in the late Triassic and quickly evolved to a large size, with species exceeding 9 m in length by the early Jurassic. All theropods were predators, and the group includes the Cretaceous dinosaur, *Tyrannosaurus rex*. Theropods evolved into birds during the Jurassic.

Ornithischians, or bird-hipped dinosaurs, also evolved from therapods during the Jurassic. They were all herbivores, and included armored forms such as *Triceratops* and the stegosaurs, as well as the highly specialized iguanodonts and duck-billed dinosaurs. These dinosaurs had highly modified teeth and jaws that allowed them to chew food thoroughly before it was swallowed, significantly shortening digestion times. Their name derives from a modification of the pelvic region, which is shown in Fig. 11.11.

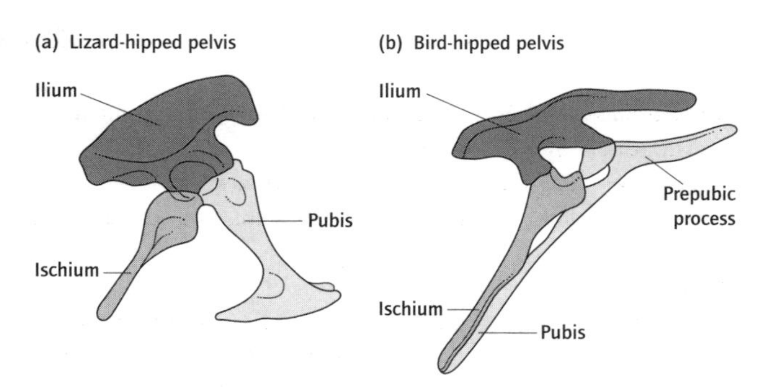

Fig. 11.11 The pelvic girdles of (a) lizard-hipped, and (b) bird-hipped dinosaurs.

Sauropods also evolved from theropods and share with them the more primitive lizard-hipped pelvic pattern. They were vegetarians like the bird-hipped dinosaurs, but evolved a different strategy for aquiring and processing the tough vegetation of the Mesozoic. These are the group of dinosaurs that evolved to extremely large size, often in excess of 20 m in length and 50 tonnes in weight.

Dinosaurs are found worldwide, including from high latitudes, where it would have been dark in winter for several months. Even in the warm global conditions of the Mesozoic, this would have necessitated a warm-blooded habit. In addition, dinosaurs were active predators and herd-dwellers, and the predator : prey ratios of dinosaurs are consistent with endothermy. Evidence of long-term protection of the eggs and young, and complicated adaptations for display and communication, suggest that dinosaurs also evolved complicated social patterns in a wide variety of different environments.

The reason for the extinction of the major diapsid groups at the end of the Cretaceous is unknown. A catastrophic meteorite impact, probably onto the Yucatan region of Mexico, may have been the cause of this mass extinction. However, there is also evidence for vast igneous eruptions in India close to the boundary that might have been responsible for significant climatic change. Most puzzling is why dinosaurs, marine reptiles, and pterosaurs perished when birds and mammals survived.

Birds

Birds evolved from theropods during the Jurassic. An intermediate step, the primitive bird *Archaeopteryx*, is known from late Jurassic rocks of southern Germany. This species is very similar in form to a small carnivorous theropod, *Deinonychus*, but it has feathers. It was probably a glider, or at best a poor flyer, suggesting that feathers evolved for another purpose than locomotion. Birds were a minor component of Jurassic and early Cretaceous ecosystems. They became more diverse and common late in the Cretaceous, and appear to have been mainly aquatic or shoreline species, including diving birds. As with pterosaurs, this may be an artifact of preservation.

Large flightless birds evolved early in the Cenozoic. This group, known as ratites, grew up to 3 m in height and included some of the top predators of the early Cenozoic. Their descendents today include ostriches and emus. Perching song birds, which dominate modern faunas, evolved in the Miocene. It may be that this was the time when birds evolved into a tree-dwelling habit and colonized dense woodland.

Glossary

Acanthodian – lightly armored, jawed fish characterized by fins supported by a frontal spine.
Acheulian – tool suite associated with *Homo erectus.*
Actinopterygian – ray-finned, bony fish, including most modern fish.
Agnathan – jawless fish.
Anapsid – primitive reptiles represented by modern turtles and tortoises. There are no holes in the skull behind the eye.
Articular – lower jaw bone in reptiles that articulates with the upper jaw, and an ear bone in mammals.
Chondrichthyan – group of fish characterized by a cartilaginous skeleton, represented by modern sharks and rays.
Creodontids – early group of carnivorous mammals, now extinct.
Cynodont – mammal-like reptiles with many mammalian characteristics, including whiskers.
Dentary – lower jaw bone of mammals and one of the lower jaw bones of reptiles.
Diapsid – group including dinosaurs, birds, marine reptiles, modern reptiles, and pterosaurs, characterized by two skull apertures behind the eye.
Gnathostome – jawed fish.
Hominid – group of species including *Paranthropus*, *Australopithecus*, and *Homo* that includes our direct ancestors and no other living group.
Ilium – large, dorsal, blade-like bone of the pelvis.
Ischium – rear-facing bone of the pelvis.
Marsupial – mammal that broods live young in a pouch.
Masseter – large chewing muscle of mammals.
Monotreme – mammal that lays eggs.
Mousterian – tool suite associated with Neanderthal man.
Oldowan – tool suite associated with *Homo habilis.*
Ornithischian – bird-hipped dinosaurs.
Osteichthyan – bony fish.
Pelycosaur – primitive group of mammal-like reptiles.
Placental – mammal that has a long gestation period and gives birth to large, live young.
Placoderm – heavily armored, jawed fish, common in the Devonian.
Prepubic process – forward-facing pelvic bone characteristic of bird-hipped dinosaurs.
Pubis – pelvic bone that faces forwards in lizard-hipped dinosaurs and backwards in bird-hipped dinosaurs.
Quadrate – upper jaw bone of reptiles that articulates with the lower jaw, and an ear bone in mammals.
Ratites – large, predatory flightless birds characteristic of the Palaeocene.
Rhipidistian – extinct group of lobe-finned fish that was probably the ancestor of tetrapods.
Sarcopterygian – lobe-finned fish, including lungfish, coelocanths, and rhiphidistians.
Sauropod – large herbivorous dinosaurs.
Secondary palate – division in mammalian skulls that allows the animal to eat and breath at the same time.
Squamosal – upper jaw bone that articulates with the lower jaw in mammals, but not in reptiles.
Synapsid – group including modern mammals and mammal-like reptiles, characterized by a single skull aperture behind the eye.
Therapsid – advanced group of mammal-like reptiles, specialized for temperate and high latitudes.
Theropod – carnivorous dinosaurs.

12 Land plants

- Land plants probably originated from green algae. Current evidence suggests that nonvascular plants evolved during the Silurian but the earliest uncontroversial fossils of nonvascular plants are from the Lower Devonian.
- With the development of vascular tissue, plants became increasingly independent of aquatic habitats.
- The Devonian was a period of rapid plant diversification, characterized by the appearance of seed-bearing plants and the first forests.
- Mesozoic terrestrial ecosystems were dominated by conifers, cycads, and ferns.
- Flowering plants diversified spectacularly in the Cretaceous and are the most abundant group of plants.

Introduction

The fossil record of plants is typically fragmentary. Generally plants have a low preservation potential and assemblages are usually composed entirely of disarticulated material. During their life cycle plants may shed some of their component parts. Other plants become broken up or fragmented due to taphonomic processes. As a result, separate parts of the same plant are often assigned to different organ genera (Fig. 12.1). The extent of plant preservation is dependent on the morphology and biology of the plant and different plant parts have different preservation potentials.

Plant material is often preserved when the pressure of accumulated sediments compresses it, resulting in the removal of the soluble plant components and reducing the material to a thin carbon film. These compression fossils are common in nonmarine, deltaic environments. Further modification may result in the complete loss of organic material and only an impression of the plant may remain. Plant fossils, therefore, may have an impression and a compression surface. The way in which the rock splits determines the type of fossil. Plant material may also be preserved as molds or casts. More rarely, plant material is preserved through perimineralization or petrification. Mineral-saturated fluids infiltrate the cells and intercellular spaces. Subsequent crystallization reveals the internal structure of the plant parts.

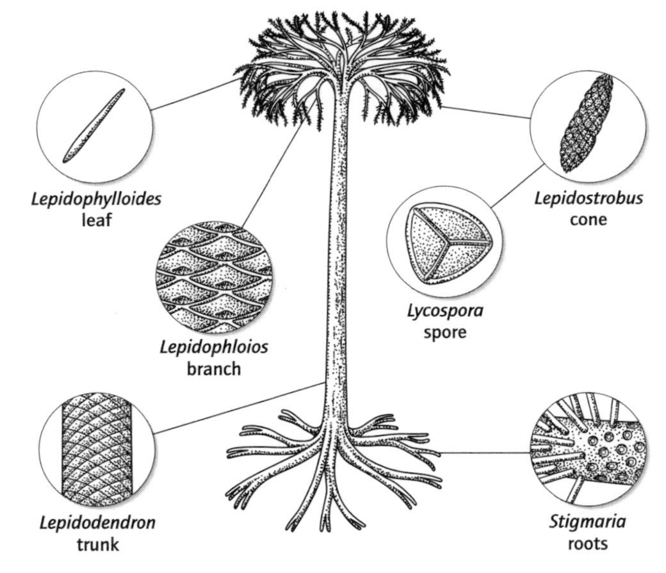

Fig. 12.1 *Lepidodendron* reconstruction (height 50 m).

Plant classification

A classification of land plants is given in Table 12.1. Informal groupings are used in this scheme. As an artificial classification it provides a working description of plant diversity rather than an explanation of the evolutionary relationships. Some of the groups are unnatural, for example "seed ferns" incorporates the seed plants that are not included in the other groups.

Land plants are separated into those with a vascular system and those without. There are three important groups of nonvascular plants all of which are extant: the hornworts, liverworts, and mosses. Vascular plants are separated into seedless and seed plants. Seedless vascular plants include three extinct groups, and three groups with living relatives: club mosses, ferns, and horsetails. Vascular seeded plants are divided into plants with naked seeds, the gymnosperms, and plants with seeds enclosed by a fruit, the angiosperms. The most important groups of gymnosperms are the conifers, cycads, and ginkgoes. Angiosperms are the flowering plants, the dominant living flora.

Table 12.1 Plant classification, based on common morphological characteristics and not common lineage. These are the most useful groupings for fossil plants.

	Description	Example	Range
Nonvascular plants			
Anthocerophytes (hornworts)	Similar to liverworts but sporangia are able grow continuously		Early Devonian–Recent
Hepatophytes (liverworts)	Small plants with a flattened leaf-like body (thallus)		Late Devonian–Recent
Bryophytes (mosses)	Filamentous mat with simple leaves and root-like structures (rhizoids)		Carboniferous–Recent
Vascular seedless plants			
Rhyniophytes (*Cooksonia*)	Extinct plants with dichotomously branched, simple axes. Sporangia terminal		Silurian–early Devonian
Lycophytes (club mosses)	Leafy plants with sporangia on upper surface of leaves or at the leaf–stem intersect		Late Silurian–Recent
Zosterophylls (*Sawdonia*)	Extinct dichotomously branched plants, sometimes with spiny axes. Sporangia on side of axes		Early–late Devonian
Progymnosperms (*Archaeopteris*)	Extinct fern-like plants with woody tissue		Mid–Upper Devonian

Table 12.1 *(cont'd)*

	Description	Example	Range
Vascular seeded plants			
Pteridophytes (ferns)	Large-leaved plants with sporangia on the lower surface		Mid-Devonian–Recent
Sphenophytes (horsetails)	Plants with leaves and branches fused in whorls. Fertile branches have terminal cones		Late Devonian–Recent
Vascular seeded plants – gymnosperms			
Seed ferns	Fern-like plants with spores on the leaves		Late Devonian–Jurassic
Conifers	Woody trees with needle or scale-like leaves. Seeds in cones		Early Carboniferous–Recent
Cycads	Woody, stemmed plants with palm- or fern-like leaves and cones		Early Carboniferous–Recent
Bennettitales (*Cycadeoidea*)	Extinct plants that resemble cycads but have flower-like cones		Triassic–late Cretaceous
Ginkgoales (ginkgoes)	Woody trees with fan-shaped leaves. Cones absent		Late Triassic–Recent
Gnetales (gnetae)	Unusual group with cone clusters resembling flowers	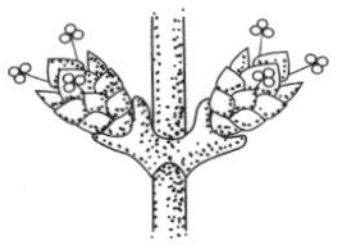	Late Triassic–Recent
Vascular seeded plants – angiosperms			
Angiosperms (flowering plants)	Plants with flowers. Seeds enclosed within a fruit		Cretaceous–Recent

Plant life histories

Evidence suggests that the alternation of generations is an important feature in plant evolution. Two morphologically distinct vegetative phases can exist in plants: the gametophyte and sporophyte stages (Fig. 12.2). The gametophyte generation is a thin and filmy thallus prone to desiccation. It is composed of haploid cells and produces the sex cells, the gametes. Gametes fuse to form diploid zygotes. The sporophyte generation develops from the diploid zygote and is the spore-producing stage. Germination of spores produces the next gametophyte generation. Sporophytes have well-developed cuticles and are adapted to terrestrial environments.

Different generations have dominated the evolution of land flora. In the most primitive plants, the gametophyte generation is dominant. This is also the situation in modern mosses. Evolution in the pteridophytes (ferns), gymnosperms, and angiosperms is characterized by the increasing dominance of the sporophyte.

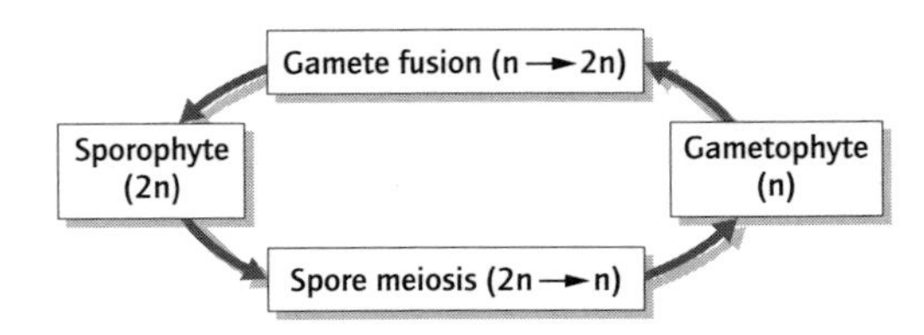

Fig. 12.2 Alternation of generations in plants.

Key steps in plant evolution

1 *Colonization of the land*: With the evolution of several new morphological features (roots, stomata, cuticles, lignin, leaves, and vascular tissues) and the development of new reproductive structures (spores, pollen, or seeds) plants became independent of the aquatic environment. On land resources are spatially separated. Underground root systems absorb water and minerals whilst the stem and leaves generate organic products through photosynthesis. Vascular tissues transport nutrients and water to different parts of the plant enabling plants to thrive in the terrestrial environment.

2 *Evolution of leaves*: Leaves are the main photosynthetic organs of most living plants. Early, leafless vascular plants probably photosynthesized using the stem. Two types of leaves have evolved: small and simple, and large and divided. Small simple leaves that originate directly from the stem and have a pronounced mid-vein may have been derived from stem outgrowths (spines). Large, flattened, divided leaves may have originated from the repeated subdivision of branch tips and the development of photosynthetic tissue (webbing) between the branches.

3 *Origin of the seed*: The dominance of the sporophyte with its greater terrestrial adaptation and the reduction of the gametophyte to a few haploid cells within a heavily protected seed has allowed gymnosperms and angiosperms to colonize a much wider range of drier habitats.

4 *Development of forests*: Woody plants undergo a secondary growth that increases the width of roots and stems. This secondary growth is produced from a unique meristematic tissue, the cambium, which produces xylem internally and phloem externally. The xylem is thickened with lignin, making up the secondary wood that enables plants to grow taller. The first forests are known from the late Devonian. Forests are vertically tiered systems with many different habitats. Vascular plants underwent a dramatic diversification in response to the creation of new habitats associated with forest development.

5 *Evolution of flowers*: Flowers have evolved from modified leaves that have assumed a reproductive function. Increased efficiency in angiosperm reproduction has been brought about by the development of insect pollination and the complex coevolutionary relationships between plants and insects. In the most extreme cases, flower structures will only allow specific insects to pollinate them.

The earliest land plants

Plausible fossil green algae, dating back to 900 Ma, have been described from the Bitter Springs Chert, a siliceous deposit from Australia. Green algae may have colonized shallow water and shoreline habitats subject to periods of exposure. In this way algae living in marginal environments would have become adapted to periods of exposure and therefore life on land. Such plants would have had a selective advantage over plants living continually submerged.

The first land plants were probably nonvascular, that is they did not have a specialized system for transporting water and nutrients throughout the plant body. Nonvascular plants form three separate groups: hornworts (anthocerophytes), liverworts (hepatophytes), and mosses (bryophytes) (see Table 12.1). Due to their low preservation potential their fossil record is fragmentary. Cladistic analysis implies a Silurian origin for nonvascular plants. The oldest fossilized nonvascular plants are liverworts described from the Lower Devonian of Belgium (Fig. 12.3). The reconstruction of this plant shows numerous, erect, slender stems with terminal sporangia. Liverworts are also known from the Devonian of New York State, USA.

Simple vascular plants are known from the Silurian. The plant bodies of these early forms were slender, leafless, and without roots. Sporangia were simple, swollen branch tips. The oldest true vascular plant is *Cooksonia* (Fig. 12.4), flourishing from the mid-Silurian to early Devonian times. Silurian forms are only a few millimeters tall, while incomplete Devonian forms have been found up to 6.5 cm in height.

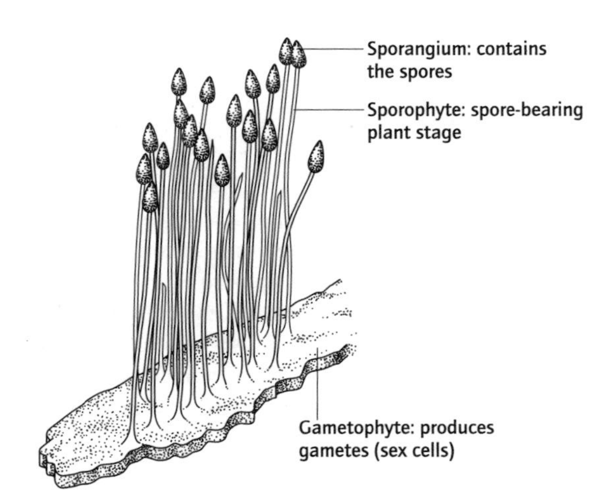

Fig. 12.3 *Sporogonites*— this fossil liverwort is the oldest, fossilized nonvascular plant (2 cm tall).

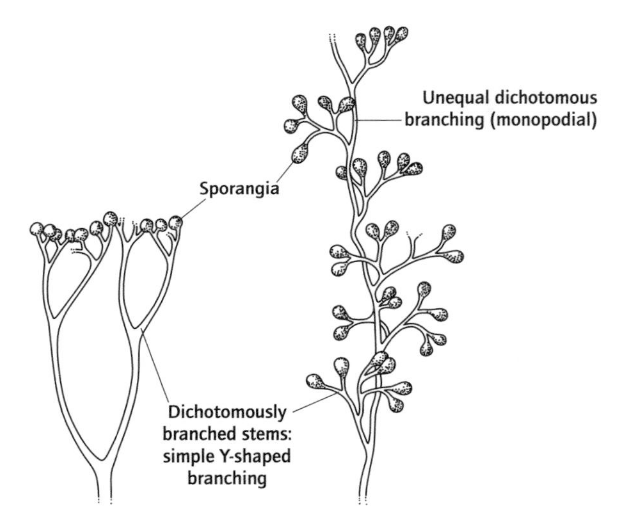

Fig. 12.4 Two species of *Cooksonia*; the taller specimen is 6.5 cm high.

Colonization of the land

The transition from water to land required a number of adaptations in plants. The main environmental challenges presented by moving from a water to an air medium are: obtaining sufficient water, limiting water loss, adapting the photosynthetic system to deal with higher light intensities and different carbon dioxide: oxygen ratios in the atmosphere, and tolerating changes in temperature. Critical terrestrial adaptations in plants may be summarized as follows:

1 *Roots*: Roots anchor plants in soil, giving plants support and providing a mechanism for obtaining nutrients and water.

2 *Cuticle*: This waxy, water-insoluble coating on stems and leaves reduces water loss by evaporation.

3 *Stomata*: Stomata are microscopic pores on plant stems and the undersurfaces of leaves that allow carbon dioxide into the plant for photosynthesis; the consequent water loss may perform a cooling function. The balance between carbon dioxide entry and water loss determines the possible habitats that land plants occupy.

4 *Vascular system*: This specialized system enables the transport of vital materials throughout the plant. Two conducting tissues – the xylem and phloem – transport water containing dissolved mineral salts and organic nutrients, respectively.

5 *Lignin*: Lignin is a strengthening polymer found in vascular plants that provides support and enables the plant to stand erect. Lignin also enables plants to grow taller.

6 *Leaves*: Leaves contain chloroplasts that are adapted to higher terrestrial light intensities and carbon dioxide levels, and have an increased surface area for photosynthesis.

7 *Spores, pollen, and seeds*: Land plants developed spores, pollen, and seeds that were resistant to desiccation and could be transported by the wind.

Early vascular plants

The transition from aquatic to terrestrial habitats was gradual taking place over tens of millions of years. Plants steadily adapted to the environmental challenges of the land and became less dependent on aquatic habitats.

The first true vascular plants are Middle Silurian, although resistant, cutinized spores are known from Upper Ordovician sediments suggesting that they may have evolved earlier. Known from the mid-Silurian, early leafless and/or rootless forms do not have all the features associated with the vascular system. The oldest, simple vascular tissues are late Silurian, and stomata are first recorded in the early Devonian.

The earliest vascular plants were rhyniophytes. Rhyniophytes are slender, dichotomously branched plants without leaves or roots (Fig. 12.5). *Cooksonia* is a member of this group. Coexisting with the rhyniophytes in the early and mid-Devonian were the zosterophylls. Similar to rhyniopyhtes, some members of this group had small spine-like projections (Fig. 12.6). Most of these primitive plants were very short (only a few centimeters) and were anchored by rhizomes, a horizontal underground stem. Sediments show that early vascular plants lived in wet, marshy environments and were still associated with aquatic habitats.

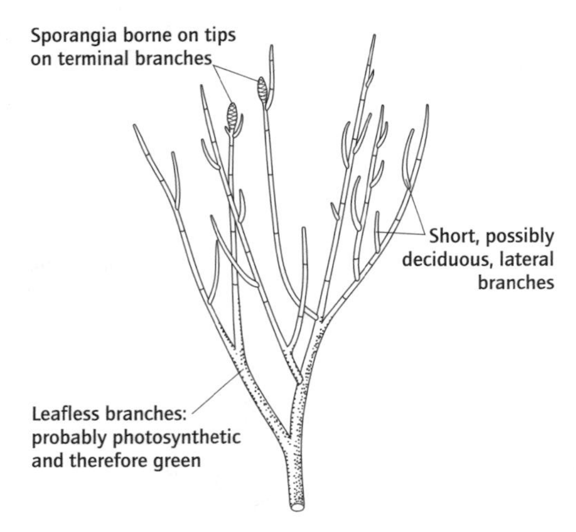

Fig. 12.5 *Rhynia* (height 17 cm).

Rhynie Chert

Complete rhyniophytes preserved in exquisite detail, as a result of siliceous perimineralization, are known from the Lower Devonian Rhynie Chert of northeast Scotland, UK. Material from this locality shows that the vascular system of early land plants was very simple, constructed only of water-conducting xylem. Stomata preserved in the epidermis provide evidence that these plants photosynthesized and it is presumed that the stems were green. All the plants are without roots. *Rhynia* is anchored by tuft-like rhizomes, whilst *Horneophyton* has bulb-like structures with small thread-like rhizomes (Fig. 12.7).

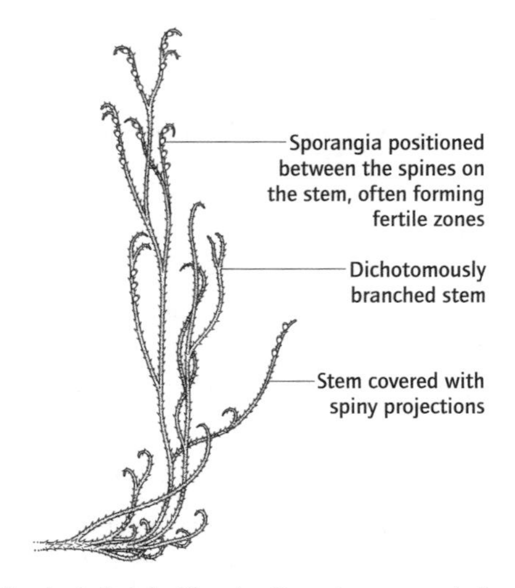

Fig. 12.6 *Sawdonia* (height 20 cm), a Devonian zosterophyll.

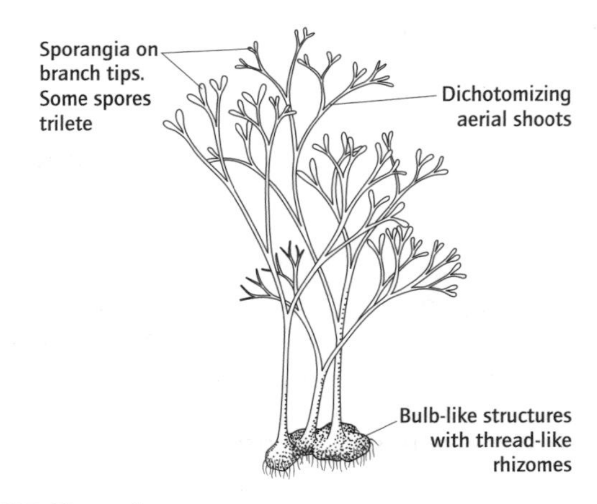

Fig. 12.7 *Horneophyton*.

Spore-bearing plants

Advanced spore-bearing plants with true leaves and roots evolved during the Devonian: lycopods (club mosses), sphenophytes (horsetails), pteridophytes (ferns), and progymnosperms (precursors to the seed-bearing gymnosperms). Some plants developed specialized woody tissue enabling them to attain the stature of trees.

Lycopods (lycophytes) formed a major part of the Devonian flora. Two distinct evolutionary lines developed from the lycophytes. One line, now extinct, evolved into the tall trees that dominated the Carboniferous coal swamps. The second group remained small and did not develop woody tissue. Living lycopods are generally low-lying plants with small, scale-like leaves. These small leaves probably represent modified stem outgrowths. Lycopod sporangia are positioned at leaf–branch intersections and are clustered in protective cones.

Horsetails (sphenophytes) have a jointed hollow stem and distinctive spiked leaves fused in whorls. Similar to lycopods, the sporangia are organized in terminal cones. Fossil horsetails tended to be much larger than modern forms. Some Carboniferous species grew up to 20 m in height.

Ferns (pteridophytes) are known from the Carboniferous and are the most common, living, spore-bearing plants. Sporangia are situated on the underside of large compound leaves, called fronds, formed of many leaflets. Each frond has a branched vein system. Such leaves probably evolved through the formation of webbing between branch tips. Most living ferns have leaves that are produced from a rhizome at ground level. Some fossil forms and tropical tree ferns have an upright trunk-like stem many meters tall.

The progymnosperms were the forerunners of the gymnosperms, seed-bearing plants. Superficially similar to tree ferns, these plants have woody trunks with a structure similar to some conifers. This gives them a vegetative structure similar to seed plants and reproductive mechanisms similar to ferns. Progymnosperms are only known from Devonian and Lower Carboniferous rocks.

Carboniferous coal forests

Immense forests dominated by spore-bearing plants thrived in low-lying, swampy areas during the Carboniferous (Fig. 12.8). Extremely tall club mosses, *Lepidodendron*, and *Sigillaria* dominated the floodplain vegetation. *Lepidodendron* had a tall, unbranched trunk with a small canopy of branches at the top. Some plants exceeded 50 m in height and 2 m in diameter at the trunk base. Underground branched axes, *Stigmaria*, with root-like appendages, supported the massive trunk. The giant horsetail *Calamites* also occupied low-lying, boggy areas, although it was much rarer. This tree-like plant reached 30 m in height and the trunk exceeded 40 cm in diameter. Ferns and conifers colonized drier, more elevated areas. Towards the end of the Carboniferous the low-lying marshy areas dried out as the climate became more arid. The early seed-bearing coniferophyte, *Cordaites*, was an important component of Carboniferous forests. Reaching 30 m in height this repeatedly branched tree had long simple leaves and seed-bearing cones. By the end of the Permian most club mosses, horsetails, and coniferophytes were replaced with species more suited to drier habitats.

Fig. 12.8 Reconstruction of a Carboniferous forest.

Seed-bearing plants – gymnosperms

Seed-bearing plants are divided into two groups. Those with exposed (naked) seeds, the gymnosperms, and plants that flower and produce seeds within a fruit, the angiosperms. Plants with seeds first appeared in late Devonian times and proliferated during the Upper Palaeozoic. Gymnosperm development peaked in the Mesozoic. Seeds have four main reproductive advantages over spores:

1 A multicellular embryonic plant is held within the seed, whereas a spore is a single cell.
2 Seeds contain a food supply that nourishes the plant until it is self-sufficient.
3 Seeds have a resistant, protective coat.
4 Fertilization and pollination are independent of free water.

With the evolution of seeds, plants were no longer restricted to damp habitats for reproduction and the potential for dispersal became much greater.

Seed ferns flourished in the Carboniferous and Permian. Although superficially similar to spore-bearing ferns their internal structure is very different. Seed ferns probably developed from progymnosperms (Fig. 12.9).

The origin of conifers is uncertain but they may have evolved from seed ferns or cordaitean plants. Occupying mainly dry environments, conifers were important plants in the Carboniferous and the Permian. At 30 m in height and with long strap-like leaves, *Cordaites*, was a very distinctive Upper Palaeozoic coniferophyte (Fig. 12.10).

Modern conifers can be traced back to the Triassic when the group underwent a major radiation. Cycads and bennettitaleans were important components of the Mesozoic vegetation although there are only a few living cycad genera. Similar to ferns and palms, cycads have large, compound leaves. The stems or trunks are usually unbranched, and covered with scale-like leaf bases. Each "scale" represents a former leaf attachment site. Cycadeoids are very similar to cycads but they have different cone structures and the leaf traces are not preserved on the trunk (Fig. 12.11).

Presumably originating in the Permian, ginkgoes diversified in the Mesozoic and were widely distributed. They declined in the Cenozoic and only one ginkgo species survived to the present day. Ginkgoes are large, densely branched trees with entire or bilobed leaves. Fossil ginkgo leaves closely resemble the modern foliage. As living ginkgoes are deciduous, fossil species may also have seasonally shed their leaves.

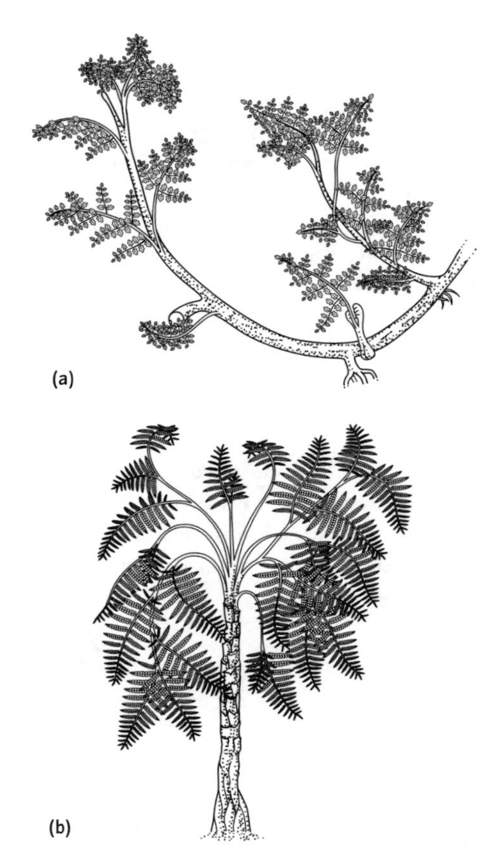

Fig. 12.9 (a) *Callistophyton*, a scrambling understorey fern (the fronds are approximately 50 cm in length). (b) *Medullosa*, a tree fern with primitive compound leaves (height 10 m).

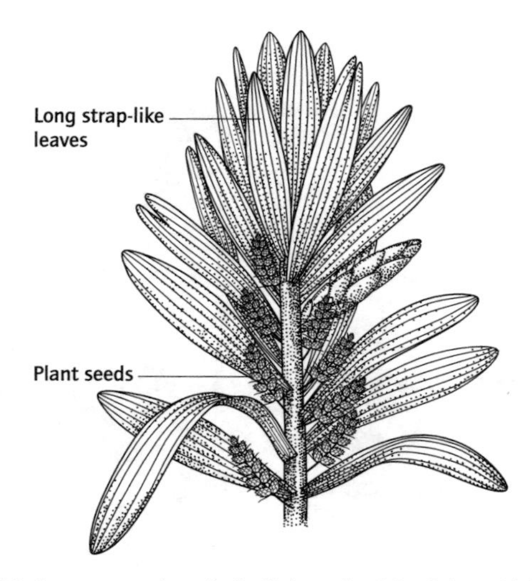

Fig. 12.10 A reconstruction of a fertile branch of *Cordaites*, with leaves and cone axes. The leaves can be up to 1 m in length.

Fig. 12.11 A reconstruction of *Cycadeoidea* (height 2 m).

Gnetales are a diverse group of gymnosperms with an increasing fossil record. The presence of flower-like cone clusters suggests that they might share a common ancestor with the angiosperms (Fig. 12.12).

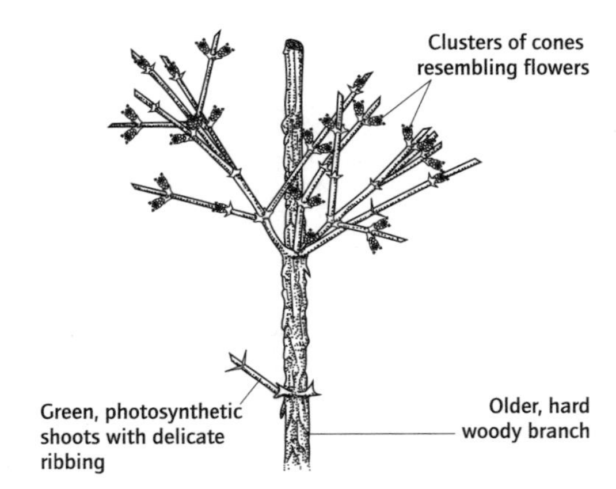

Fig. 12.12 An *Ephedra* branch with shoots (height 10 cm).

Seed-bearing plants – angiosperms

Angiosperms are the most diverse and widespread group of living plants (Fig. 12.13). Leaf impressions of angiosperm-like plants are known from the Triassic. The first true angiosperm fossils are Cretaceous (Fig. 12.14). During this period angiosperms diversified rapidly, particularly in low latitudes, and dominated most habitats by the end of the Cretaceous. Paralleling the rise of the angiosperms, spore-bearing plants and gymnosperms declined through the Cretaceous. Abundance and diversity decreased in most other plant groups; although conifer diversity remained relatively stable they were increasingly polarized, geographically, into marginal (high altitude) environments.

Angiosperms reproduce sexually using flowers. Flowers are essentially clusters of modified leaves, some with a reproductive role. There are over 235,000 living angiosperm species compared with 720 living species of gymnosperms. Angiosperm success is attributed to the following evolutionary developments:

1 A vascular tissue with vessels that provides support and transports water more effectively.
2 Diverse pollination mechanisms including the pollination of flowers by insects. These are more efficient systems than random wind pollination.
3 An enhanced reproductive process, including double fertilization, resulting in a plant embryo and a food source contained within the seed.
4 The enclosure of developing seeds within a fruit that protects the seeds and also aids seed dispersal.
5 The enclosure of the gameophyte generation within the sporophyte seed enabling angiosperms to colonize a much wider range of habitats.

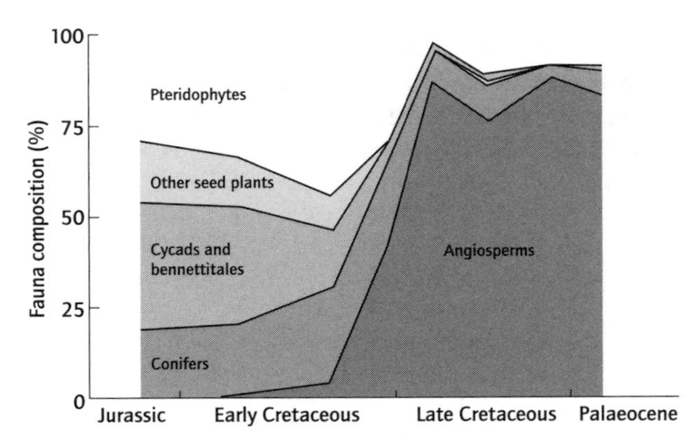

Fig. 12.13 The percentage contribution of major plant groups to the Jurassic, Cretaceous, and Palaeocene faunas.

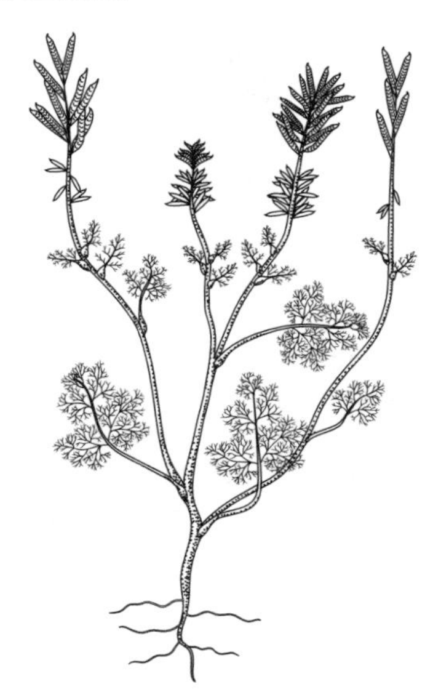

Fig. 12.14 A reconstruction of *Archaefrutus*. (Reprinted with permission from Sun *et al.* (2002) Science, 296, 899–904, fig. 3. © 2003 American Association for the Advancement of Science.)

Calamites

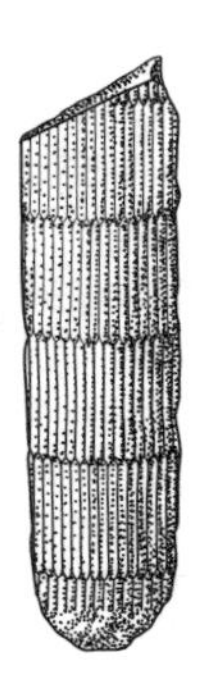

Sphenophyte

Carboniferous

Calamites (a sphenophyte) is the stem remains of the extinct giant horsetail, although it is commonly used to describe the entire plant.

Calamites is usually preserved as a cast. The outer part of the stem was made of resistant wood but the center was formed of softer tissue, pith. After death, the pith decayed rapidly leaving a hollow, wooden cylinder. Sediment then infilled the cavity, hardened, and formed a "pith cast". The texture of the outer surface of the fossil reveals the structure of the plant's vascular system on the inside of its stem.

The figured stem fragment is 18 cm in length.

Annularia

Sphenophyte

Carboniferous–Permian

Annularia is foliage associated with the giant horsetail *Calamites* (a sphenophyte). Slender leaflets were arranged in whorls around the smaller stems. Individual leaves were fused at their bases and had a single, unbranched vein running along the length of the leaf. Leaflets were approximately 4 cm in length.

Horsetails were abundant in low-lying, marshy areas in the Carboniferous.

Stigmaria

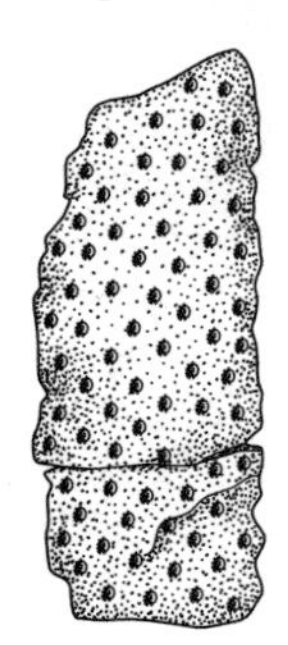

Lycophyte

Carboniferous

Stigmaria is part of the rootstock on which the massive trunk of the fossil lycopod, *Lepidodendron* (a lycophyte), rested. *Stigmaria* were horizontal, branched, underground axes with small root-like appendages. The small circular scars on the surface mark the former position of the rhizoids – small underground rootlets.

The length of the figured *Stigmaria* is 15 cm.

Sphenophyllum

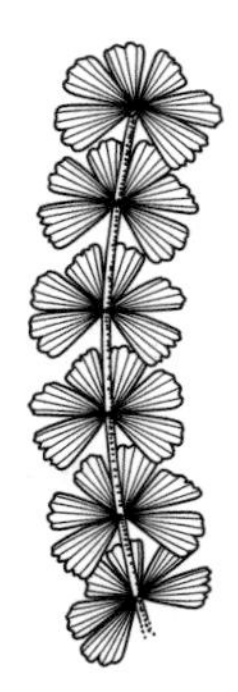

Sphenophyte

Upper Carboniferous

Foliage associated with the horsetail *Sphenophyllum* (a sphenophyte). Contemporary with the giant horsetails, these forms were similar to the modern herbaceous species, *Equisetum*.

Whorls of wedge-shaped leaflets encircled the slender stems. Leaflets had dichotomously branched veins and were approximately 1 cm in length.

Sigillaria

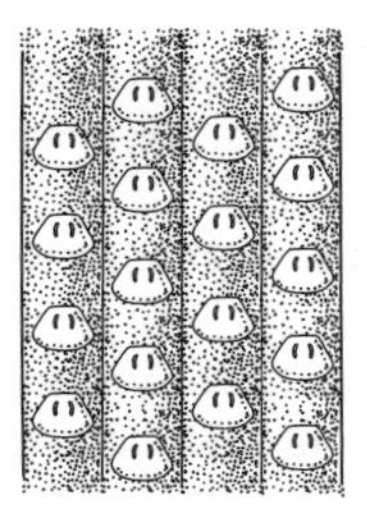

Lycophyte

Carboniferous–Permian

Part of the stem of the fossil lycopod *Sigillaria* (a lycophyte). The external texture shows the former positions of the leaf bases. Sporangia were borne on the stem surface amongst them.

The figured stem fragment is 6 cm in length.

Lepidodendron

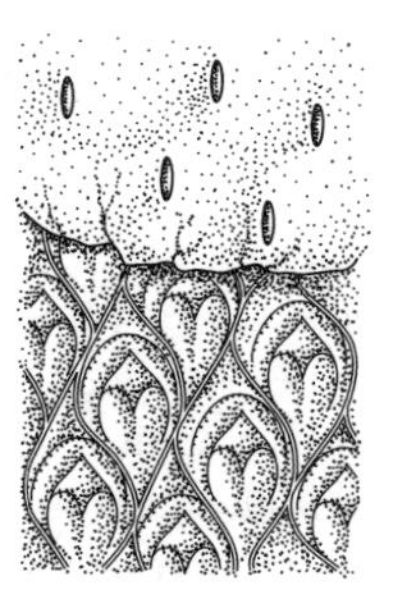

Lycophyte

Upper Carboniferous

Section of a branch showing the characteristic pattern of *Lepidodendron* leaf bases. Leaves of *Lepidodendron* were linear with a swollen attachment area. Leaf attachments sites leave a distinctive, rhombic impression (leaf cushions) on the stem surface. The close spacing of these markings shows that the foliage was dense. The leaf size varied between species and leaves were mainly restricted to smaller branches.

Leaf cushions are approximately 3 cm in length.

Sphenopteris

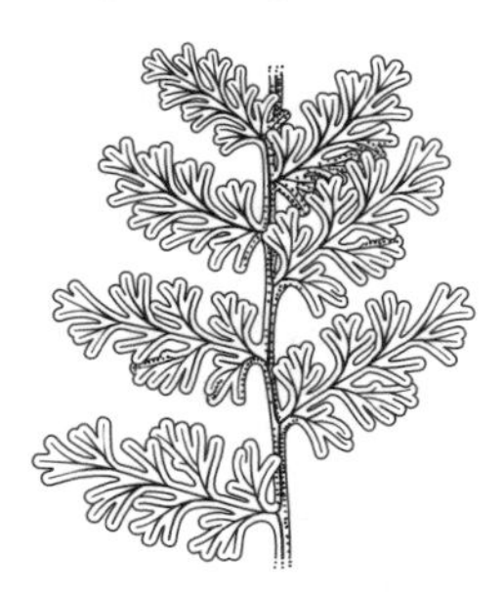

Pteridophyte

Devonian–Permian

Part of the foliage of the extinct fern *Sphenopteris* (a pteridophyte). Fronds have distinctive lobed pinnules.

The figured section of frond is 6 cm in length.

Mariopteris

Seed fern

Upper Carboniferous

Frond of the seeded fern *Mariopteris*. Fronds of *Mariopteris* were generally small (50 cm in length) and some specimens had thread-like tendrils at the leaf tips suggesting that this fern had a vine-like, scrambling habit.

Glossary

Angiosperms – flowering plants that produce seeds within a protective fruit.
Cambium – meristematic tissue found within the roots and stems of perennial plants.
Chloroplast – cell organelle responsible for photosynthesis.
Coniferophytes – early (Carboniferous) seed-bearing plants, such as *Cordaites.*
Cuticle – waxy coating that reduces water loss.
Dichotomously branched – axes divided into two equal parts. A primitive form of branching common in early vascular plants, e.g., *Rhynia.*
Diploid cell – vegetative cell of the sporophyte generation (chromosome number 2n).
Double fertilization – takes place in the developing seed and results in the formation of a seed with a food source (endosperm).
Gametes – haploid sex cells that fuse during reproduction to form a diploid zygote which develops into the sporophyte.
Gametophyte generation – haploid vegetative gamete-producing stage in the life history of a plant. Gametes fuse and grow into sporophytes.
Gymnosperm – vascular seed plant with exposed (naked) seeds that are usually borne within cones.
Haploid cell – cell of the gametophyte generation (chromosome number n).
Lignin – complex polymer in the plant cell walls that confers rigidity and provides support in terrestrial plants, particularly in woody species.
Lycopod – club moss (a lycophyte).
Meiosis – reduction division that converts diploid cells to haploid cells.
Meristem – plant tissues that remain embryonic as long as the plant lives. Its products differentiate into the tissues of the stem and root.
Monopodial branching – dichotomous branching where one axis is dominant.
Nonvascular – plants without a vascular system.
Organelle – membranous subcellular structure specialized for a particular function.
Phloem – vascular tissues that transport organic nutrients throughout plants.
Pollen – immature male gametophyte that is transferred through pollination to the female gametophyte.
Rhizoid – fine hair-like extensions, usually from rhizomes that act as absorptive organs.
Rhizome – horizontal stem at, or just below, ground level that supports vertical axes.
Sporangium (plural sporangia) – structure in which the spores develop.
Spore – haploid cell produced by the sporophyte that germinates to produce the gametophyte generation.
Sporophyte generation – diploid vegetative, spore-producing stage in the life history of a plant.
Stoma (plural stomata) – small pores that allow gas exchange between the environment and plant interior.
Thallus – haploid vegetative gametophyte body.
Trilete – spores that have been produced by meiotic cell division.
Vascular tissue system – system that transports water and organic nutrients throughout the plant.
Xylem – vascular tissues that transport water from roots to rest of the plant.
Zygote – diploid cell resulting from fusion of gametes during sexual reproduction.

13 Microfossils

- Microfossils are fossil remains of microscopic organisms or microscopic parts of macroorganisms.
- The main groups considered are protists, ostracodes, conodonts, spores, and pollen.
- Microfossils are important biostratigraphic indicators.
- Microfossils are useful in palaeoenvironmental reconstruction.
- Changing patterns of palaeoceanographic circulation can be identified using microfossils.

Introduction

Micropalaeontology is the study of microscopic fossils. Because it is based only on fossil size, the term unites groups that are otherwise unrelated and includes microscopic organisms and microscopic parts of macroorganisms. Many sediments contain microfossils and they are important biostratigraphic, palaeoenvironmental, and palaeoceanographic indicators. Microfossils are important in biostratigraphy due to their abundance, global distribution, and durability. Their sensitivity to environmental conditions makes them useful in palaeoenvironmental reconstructions.

The groups considered in this chapter are listed in Table 13.1. Protists are single-celled eukaryotic organisms. They can be divided into "plant-like" organisms that are autotrophic and "animal-like" organisms that are heterotropic. Multicelled, but small, organisms are represented by ostracodes. Microscopic parts of macroorganisms covered are conodonts (tooth-like elements of vertebrates) and spores and pollen (the reproductive parts of plants).

Table 13.1 The microfossil groups covered in this chapter.

Main groups of microfossils	Fossil	Description	Size	Composition	Habitat	Range
Plant-like protists	Acritarchs	Hollow vesicles, probably algal cysts	< 100 μm	Organic	Marine	Proterozoic–Tertiary
	Dinoflagellates	Algae with organic-walled cysts	5–150 μm	Organic	Aquatic	Silurian–Recent
	Coccolithophores	Algae with spherical shells formed by platelets	< 50 μm	Calcite	Marine	Triassic–Recent
	Diatoms	Algae contained within a frustule comprised of two elongate or circular valves	< 200 μm	Silica	Aquatic	Jurassic–Recent
Animal-like protists	Radiolaria	Protists with a delicate exoskeleton in the form of a spherical or conical mesh	0.03–1.5 mm	Silica	Marine	Cambrian–Recent
	Foraminifera	Testate protists with single-chambered flask-like or complex multichambered tests	0.01–100 mm	Organic Agglutinated Calcareous	Marine	Cambrian–Recent
Microinvertebrates	Ostracodes	Crustacean arthropods with a bivalved carapace	1–10 mm	Calcareous	Aquatic	Cambrian–Recent
Microvertebrates	Conodonts	Tooth-like elements of a jawless vertebrate	0.1–5 mm	Phosphatic	Marine	Cambrian–Triassic
Plants	Spores	Resistant part of a plant	5 μm–4 mm	Organic	Terrestrial	Devonian–Recent
	Pollen	Resistant part of a plant	5–200 μm	Organic	Terrestrial	Devonian–Recent

Plant-like protists

Acritarchs

Acritarchs are hollow, organic-walled microfossils. They are believed to represent cyst stages in the life cycles of planktonic algae similar to a modern group, dinoflagellates, because both groups produce a characteristic molecule, dinosterane. They are one of the oldest groups of fossils and underwent a major radiation in the late Precambrian.

Morphology

Most acritarch vesicles range between 50 and 100 μm in size and are usually preserved as compressed films in black shales. Acritarchs are generally spheroidal but their shape is very variable. Vesicle walls may be single or double layered and most central chambers have an opening considered to be for the release of the motile stage. Externally, acritarchs may be smooth or granulated and most have processes projecting from the vesicle surface. Some processes are branched or have elaborate distal structures, supported by stiff buttresses, others are more simple and flexible (Fig. 13.1). Acritarchs are classified on the basis of shape, wall structure and thickness, structure of opening, ornamentation, and form of the processes.

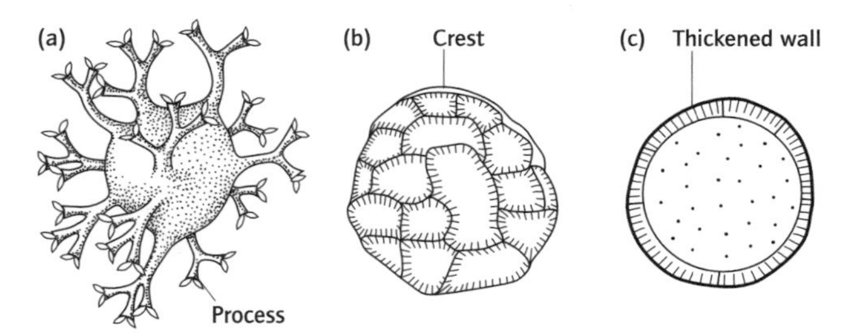

Fig. 13.1 Acritarch morphology: (a) subcircular vesicle with broad-based processes that divide into two distally (diameter 30 μm); (b) subcircular vesicle with cristate (crested) surface sculpture (diameter 40 μm); and (c) circular vesicle with a thickened outer wall (diameter 25 μm).

Palaeoecology

Acritarchs mainly occur in marine sediments, often in association with other marine fossils. They have a worldwide distribution and occur in large numbers, consistent with being primary producers rather than consumers. Such evidence strongly suggests that they were members of the phytoplankton. Furthermore, their morphology shows adaptations consistent with a planktonic mode of life. In common with modern phytoplankton, near-shore acritarch assemblages tend to be of low diversity, often dominated by one species, whilst offshore assemblages are more diverse.

Evolutionary history

Acritarchs are amongst the oldest documented fossils. Known from chemical fossils from 2.7 billion years ago, they first became abundant 1 billion years ago and are arguably the most complex Precambrian microfossils. Most forms are large for single cells (between 50 and 100 μm), and include species with a double-walled structure and ornate processes. Such diversity may represent an increase in marine productivity in the late Precambrian. A second radiation occurred in the early Cambrian. However, these forms are much smaller than those of the Precambrian. Acritarchs continued to flourish through the Ordovician. They were affected by the end-Ordovician extinction event but recovered in the Silurian, where they may have reached their highest diversity. This level of diversity was maintained through the Devonian period until late Devonian times when there was a distinct diversity decrease. Acritarchs remained scarce for the rest of the Palaeozoic. A few specialized forms appeared in the Permian but dinoflagellate cysts, spores, and pollen are the dominant organic-walled microfossils of the Mesozoic and Cenozoic.

The first radiation of acritarchs, in late Proterozoic times, may represent an evolutionary phase of early experimentation of eukaryotic phytoplankton. This rapid rise in phytoplankton may also be linked to the establishment of sexual reproduction, if the acritarch cysts were part of a sexual life cycle. The second radiation in the early Cambrian corresponds with the major expansion of suspension-feeders, emphasizing the important role of acritarchs in evolutionary history.

Dinoflagellates

Dinoflagellates are aquatic, unicellular organisms with organic-walled cysts. Usually referred to as algae, dinoflagellates have both plant-like and animal-like characteristics. During their life history about 10% of dinoflagellates develop resistant cysts that readily fossilize (Fig. 13.2). These cysts are first known from the Silurian and are important biostratigraphic indicators.

Morphology

The dinoflagellate life cycle has two stages: a motile stage, that rarely fossilizes, and a more durable benthic cyst stage. Cysts are formed from resistant organic material. Surface ornamentation may be smooth, granulated, or have raised crests and spines.

Ecology/palaeoecology

Most living dinoflagellate species are photosynthetic and marine. They form an important part of the oceanic plankton and are one of the main primary producers in the open sea. As a group they are tolerant of a wide range of temperatures and salinities. Under some conditions shallow water dinoflagellate blooms, called red tides, can poison other marine groups and cause mass mortality.

Dinoflagellate palaeoecology is difficult to determine as only a few living species produce cysts comparable with those found in the fossil record. Furthermore, cysts are easily transported by oceanic currents and fossil cysts may be found in areas not associated with the living species. However, some fossil forms are used as indicators of palaeotemperature.

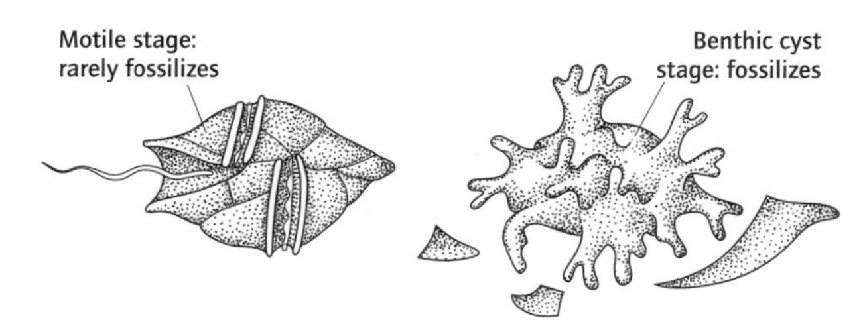

Fig. 13.2 Dinoflagellate motile and cyst stages (length approximately 25 μm).

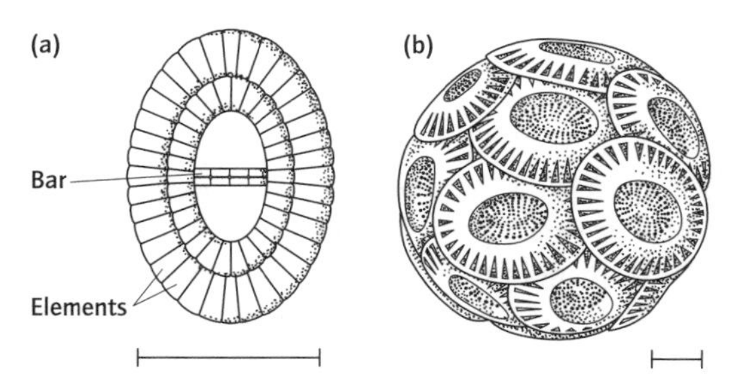

Fig. 13.3 (a) Coccolith. (b) Coccosphere and coccoliths (scale bar 1 μm).

Coccolithophores

Coccolithophores are marine, unicellular, photosynthetic plankton of extremely small size (< 50 μm and therefore termed nannoplankton). They secrete calcite platelets, coccoliths, that interlock to form a spherical shell, the coccosphere. Coccoliths have a distinctive circular or elliptical form. Similar-shaped calcite nannofossils form the Mesozoic and Cenozoic chalks.

Morphology

Coccolithophores build a spherical skeleton from between 10 and 30 wheel-shaped platelets. These are very small, typically 8 μm in diameter. Coccolith shape and structure is variable. Typically, a coccolith has an oval button-like form with a central cross bar and radially arranged elements (Fig. 13.3). Fossil platelets that are of a similar size but are pentagonal, rhombohedral, star-shaped, or horseshoe-shaped often occur in association with elliptical coccoliths. These are sometimes referred to as nannoliths.

Ecology/evolution

Living coccolithophores are restricted to the photic zone and prefer oceanic water of normal salinity between 35 and 38‰ (parts per thousand). With dinoflagellates they are one of the main primary producers in the oceans. In common with other phytoplankton, coccolithophore diversity is highest in the tropics and lowest in the high latitudes. Although subject to taphonomic processes, particularly dissolution and long-distance transportation, coccoliths are useful palaeoclimatic indicators. Some species are tolerant of a narrow temperature range and can be used to directly determine palaeotemperature, whilst relative temperatures may be estimated using ratios of warm-loving and cold-tolerant forms.

Coccoliths are known from the Upper Triassic but are very rare. Their abundance and diversity gradually increased through the Jurassic and Cretaceous, with a major radiation occurring in late Cretaceous times. Chalk is almost entirely formed from calcareous nannofossils and was deposited across vast areas at this time. Few coccoliths survived the end-Cretaceous extinction event. However, they rediversified and achieved a diversity maximum in the Eocene. Diversity has since fluctuated and is currently at its lowest point since the Cretaceous. Coccoliths are important in Mesozoic to Recent biostratigraphy.

Diatoms

Diatoms are unicellular algae with a distinctive two-part, siliceous skeleton, or frustule. Diatoms are found in virtually all marine, brackish, and freshwater environments and are even common at polar latitudes. Using a sticky secretion, they can form colonies or attach to the substrate. In modern oceans they are important primary producers. Diatom oozes form sediments in fertile waters with high concentrations of silica, nitrate, and phosphorous. These oozes lithify to form diatomites that are mined commercially for use as filtering agents and abrasives.

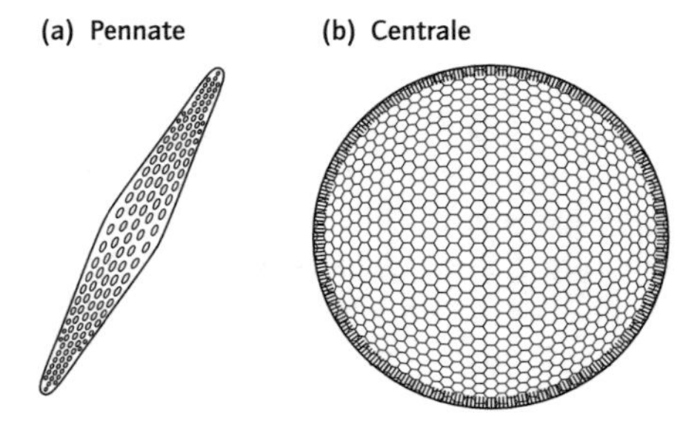

Fig. 13.4 Diatom morphology: (a) *Rhaphoneis* (approximatel 80 μm length); and (b) *Coscinodiscus* (approximately 60 μm diameter).

Morphology

The diatom frustule is formed from two overlapping, nested valves. Most frustules are between 10 and 100 μm in diameter. Valve structure and surface texture form the basis for diatom classification. Elliptical, bilaterally symmetric diatoms with a linear structural center are called pennates. Circular forms with radial symmetry around a central point are called centrales (Fig. 13.4). Ornamentation, pore pattern, and the presence of specialized structures are used to identify diatoms to genera and species level.

Ecology/palaeoecology

As diatoms are photosynthetic they are reliant on light and nutrients for growth and reproduction. They are limited to the photic zone and are most abundant in areas of oceanic upwelling, where nutrient-rich deeper water is brought up to the surface waters. Intense diatom blooms, lasting 2–3 weeks, occur in response to seasonal upwelling. Centrales are most common as marine plankton, whilst pennates are more common in benthic marine habitats or in freshwater environments.

Diatom distribution is influenced by temperature, salinity, nutrients, and pH. In the reconstruction of palaeoenvironment ratios of cold water to warm water, species have been used to estimate palaeotemperature, and interglacial/glacial stages have been identified for the Quaternary using diatoms as indicators of palaeosalinity. Diatoms can also be used to interpret water conditions in ancient lakes. As indicators of pH and fertility, diatom assemblages are also important in the monitoring of acid rain and pollution.

Evolutionary history

The first true diatoms are recorded from the Jurassic. During the Cretaceous they underwent a major radiation, and they suffered much less than other microfossil groups at the end of the period with only 23% of the genera becoming extinct. Diversity has continued to increase throughout the Cenozoic, with fluctuations in abundance being related to changing patterns of oceanographic circulation. The Palaeocene saw the first major expansion of diatoms into freshwater habitats. Diatoms reached their peak in the Miocene, possibly associated with increased volcanic activity that provided the necessary silica.

Biogeography

Areas of permanent upwelling correspond with the distribution of sediments rich in diatoms that concentrate in three main belts in the modern ocean (Fig. 13.5). A southern belt relates to the circulation of the circum-Antarctic current, an equatorial belt corresponds to the equatorial upwelling zone, and there is a weaker band in the northern oceans.

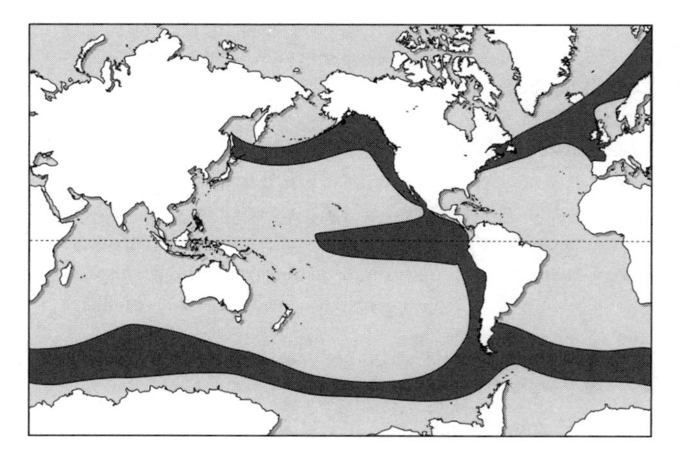

Fig. 13.5 Oceanic areas of very high silica extraction by plankton (more than 250 g of silica per square meter per year) in near-surface ocean waters. As diatoms are the dominant siliceous plankton group in the oceans this map approximates to a record of diatom production.

Animal-like protists

Radiolaria

Radiolaria are unicellular, marine microzooplankton characterized by an intricate, siliceous, internal skeleton. They are globally distributed and live at all levels of the water column. They have a long geological history, extending from the Cambrian to the present day, and their rapid evolution makes them useful biostratigraphic indicators.

Morphology

Living radiolarians may be considered as balls of protoplasm with an intricate internal skeleton, or test. Typically the test has a delicate, lattice-like structure formed from siliceous elements, including external spines, loose spicules, and internal bars. In radiolarians, buoyancy is maintained by the presence of fat globules and gas vacuoles that reduce density and by the test structure. Long spines increase drag, and the spherical and conical shapes of the tests resist sinking. Two main fossil orders are recognized on the basis of test architecture: the Spumellar and Nassellar (Fig. 13.6). Spumellarians have spherical tests that show radial symmetry, whereas nassellarians have conical, bell-shaped tests.

Ecology/palaeoecology

Of all the well-preserved protists, radiolarians have the widest oceanic distribution related, along with other zooplankton, to patterns of oceanic circulation, and species can be linked to water masses with particular characteristics. Analysis of the distribution of fossil assemblages can be used to infer the circulation patterns of ancient currents. Most species show a vertical stratification linked to temperature, and species common in deep waters in the tropics exist in shallower waters at the poles. Distinct assemblage boundaries have been identified at 50, 200, 400, 1,000 and 4,000 m, with spumellarians dominating the shallower waters and nassellarians favoring depths below 2,000 m. Radiolarians are most abundant in areas of upwelling where critical nutrients, particularly silica, are brought up to surface waters. The highest radiolarian diversity occurs at equatorial latitudes where surface waters diverge and cause upwelling. Comparisons of living radiolarians with fossil assemblages has helped identify temperature changes in water masses, for example the cooling of the Southern Ocean during the Cenozoic.

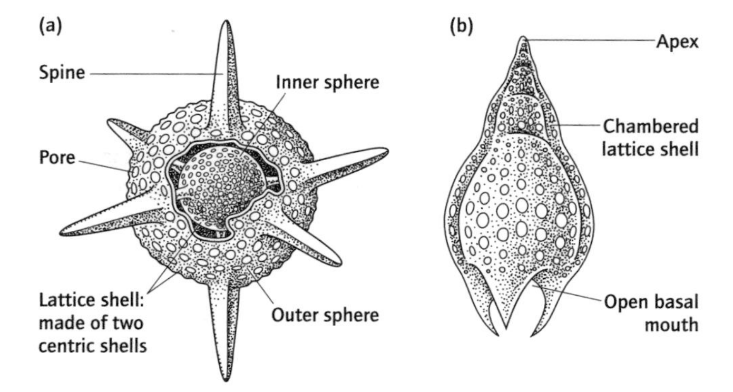

Fig. 13.6 Radiolarian morphology: (a) Spumellar (this example approximately 80 μm diameter); and (b) Nasselar (this example approximately 90 μm length).

Radiolarian oozes

As the robust siliceous test of the radiolarian is less prone to dissolution than most other microfossils, even other siliceous forms, radiolarian material accumulates in marine sediments. Below the calcite compensation depth, or the CCD, the depth at which calcite dissolves in the oceans (3,000–5,000 m), siliceous detritus dominates. In modern oceans radiolarian oozes cover 2.5% of the sea floor, mostly occurring in the deep sea sediments of the equatorial Pacific, below areas of high productivity. On average, radiolarian material accumulates at a rate of 4–5 mm per thousand years. Skeletal material may simply sink to the sea floor or tests may be contained within fecal material, which sinks more quickly and therefore has a greater chance of being preserved in sediments. Chert horizons are frequently found in the fossil record, particularly in the Mesozoic and Cenozoic, interbedded with chalk. Such nodular cherts are considered to have formed from siliceous organisms.

Radiolarians are useful for Palaeozoic and Mesozoic biostratigraphy. They are used in more recent sediments where calcareous microfossils are absent, for example where specimens have been retrieved from rocks deposited below the calcite compensation depth.

Foraminifera

Foraminifera are one of the most important fossil groups. Their distribution is global, they inhabit all marine environments, and they have a continuous fossil record from the Cambrian to the present day. Sensitive to differences in water temperature and chemistry, they are useful palaeoclimatic and palaeoceanographic indictors. They are also important for stratigraphic correlation.

Morphology

Foraminifera are unicellular organisms with an internal calcareous shell, or test. In living foraminifera most of the protoplasm is contained within the internal test. Strands of protoplasm, pseudopodia, extrude through pores in the test wall to trap food and aid movement (Fig. 13.7).

Test wall composition, structure, and shape vary and are important features in classification. Three main wall compositions have been documented: organic, agglutinated, and calcareous. Organic walls are thin, flexible membranes and agglutinated tests are constructed from cemented detritus. Practically all fossil foraminifera are calcareous. Three types of calcareous wall exist. Microgranular tests are common in Palaeozoic forms and are often recrystallized. Porcelaneous tests are translucent and hyaline tests transparent.

Test shape is determined by the arrangement and shape of the chambers and is extremely variable (Table 13.2). Most foraminiferan tests are multichambered and progressively larger chambers are secreted in simple rows or in a coil. Linear tests are formed of single (uniserial), double (biserial), or triple rows (triserial). Coiled tests vary in shape from flattened forms (planispiral) where chambers have been added in a single plane, to helical forms (trochospiral) where chambers have been added along a vertical axis. Chamber shape is also very variable. Single-chamber tests may be flask shaped, tubular, or branched. In multichamber forms chambers may be spherical or club shaped. Within the final chamber there is an opening, or aperture. Apertures may be single or multiple openings of variable shape and construction. The external surface of the test varies from completely smooth to highly ornamented. Test sculpture is considered to assist with buoyancy, anchor the foraminiferan, and deter predators.

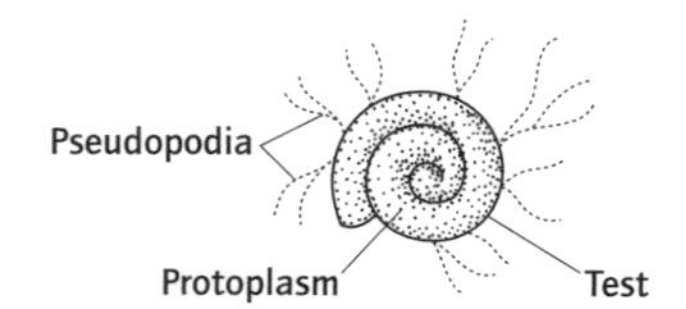

Fig. 13.7 A living foraminifera.

Table 13.2 Variation of foraminifera morphology.

Test shape	Chamber shape	Aperture type	External sculpture
Single row (uniserial)	Flask-shaped	Rounded	Spinose
Double row (biserial)	Spherical	Radiate	Costate
Coiled (planispiral)	Lunate	Slit-like	Carniate
Coiled (trochospiral)	Clavate	Sutural	Raised sutures

Ecology/palaeoecology

Foraminifera are very useful palaeoenvironmental indicators. Benthic foraminifera morphology is indicative of substrate, water depth, and seawater chemistry. Most benthic foraminifera are epifaunal and attach to the substrate using an organic membrane, the pseudopodia, or are cemented. Species associated with silty and muddy substrates tend to be thin-shelled, delicate, elongate forms. Course-grained sediments, less rich in nutrients, support sparser populations of thick-shelled, heavily sculptured foraminiferans. Furthermore, in higher energy environments, tests of free-living species are generally stronger (thicker walled). Cementing types, which prefer hard substrates, usually have a flattened or concave basal surface.

Most benthic foraminifera tolerate only normal marine salinities (approximately 35‰), and environments with fluctuating salinities are characterized by low diversity assemblages. Agglutinated foraminifera are common in marshes, whilst perforate calcareous forms favor lagoons. Hypersaline conditions are preferred only by porcellaneous foraminifera. Most foraminifera are found in marine water of normal oxygen levels, but a few thrive in low oxygen environments of the deep sea. These species are typically small and have smooth, thin-walled, calcareous or agglutinated tests.

Benthic and planktonic foraminifera assemblages are important indicators of palaeobathymetry. Indicators of palaeoenvironment and palaeodepth are summarized in Fig. 13.8.

Planktonic foraminifera are amongst the best studied groups of microfossils. They are active predators within the plankton, though some also grow photosynthetic algae within their skeleton. Their modern distribution is in five major zones that reflect modern water temperatures. These zones can be recognized through the last interglacial and glacial periods and help in the reconstruction of the oceans at warmer and cooler times than the present.

The tests of planktonic foraminifera are widely used in stable isotope analyses. These stable isotopes vary in relative proportion in the oceans in response to changing climatic features, such as temperature. The isotopic ratios of some elements are recorded by foraminifera as the animals extract material from the water to build their skeletons. Measurements made from the fossil skeletons can be used to estimate features of ancient climates such as ocean temperature, productivity, and the amount of polar ice.

Evolutionary history

The earliest known foraminifera are simple, agglutinated tubes from Lower Cambrian rocks. These forms diversified in the Ordovician and calcareous tests first occurred in the Silurian. Multichambered forms originated during the Devonian. Calcareous foraminifera continued to flourish in the Carboniferous and Permian, where a variety of test architecture with a high level of complexity existed. After the end-Palaeozoic mass extinction event benthic foraminifera underwent a major radiation adopting new habitats (lagoons, marshes, and reefs) and expanding into deeper water environments. Planktonic foraminifera first appeared in the Middle Jurassic. During the Cretaceous they underwent a major radiation, their global distribution and rapid evolution making them important zonal fossils for this period. Of the 300 species recognized for the Cretaceous, only five species of planktonic foraminifera survived the end-Cretaceous extinction event. A major radiation occurred in early Cenozoic times, although there was a severe extinction of benthic forms at the Palaeocene–Eocene boundary. Foraminifera diversified during the Eocene but suffered a further extinction in tropical taxa associated with cooling at the end of the epoch. Radiation followed in the Miocene, related to a warmer climatic phase.

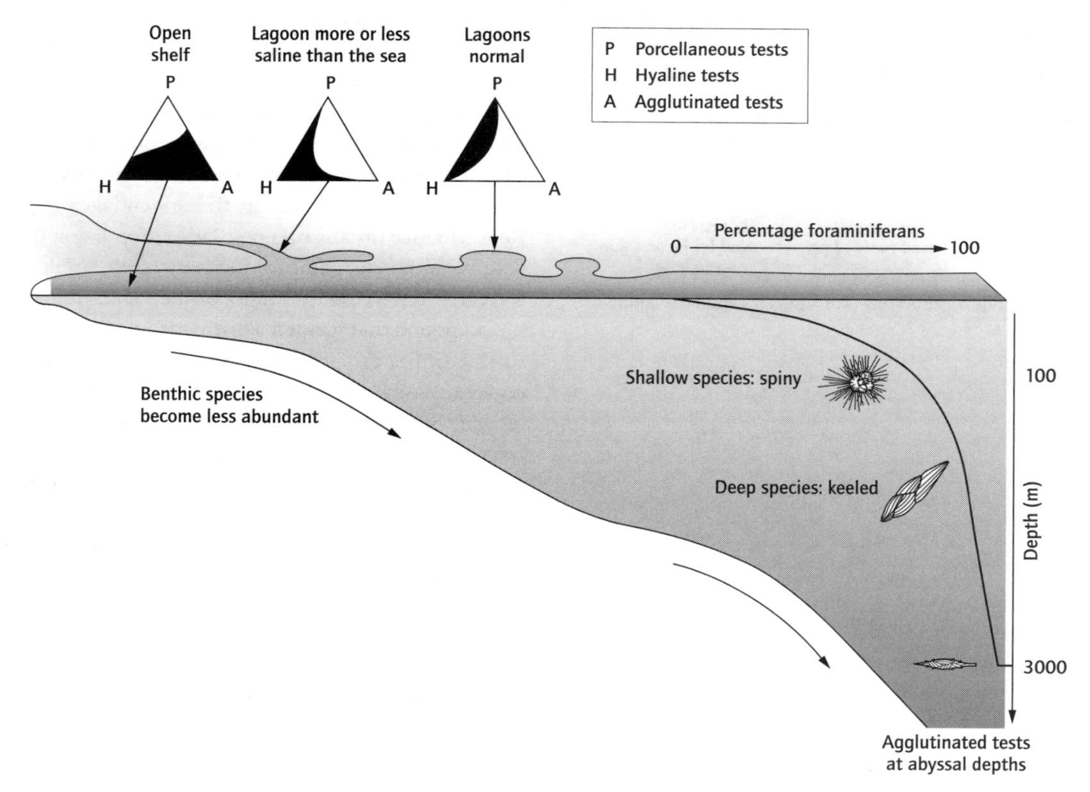

Fig. 13.8 The environmental distribution of foraminifera.

Microinvertebrates

Ostracodes

Ostracodes are small crustaceans enclosed within a bean-shaped, two-part shell. They live in all aquatic environments, have a fossil record extending from the Cambrian, and are useful indicators of past salinity.

Morphology

Living ostracodes secrete a calcareous carapace formed of two slightly overlapping, ovate, hinged valves. Most ostracodes are less than 2 mm in length, although some Palaeozoic species reached 80 mm and a few living forms are 20–30 mm in length. Carapaces can be heavily calcified and ornamented with ribs, ridges, or tubercles or can be smooth and featureless.

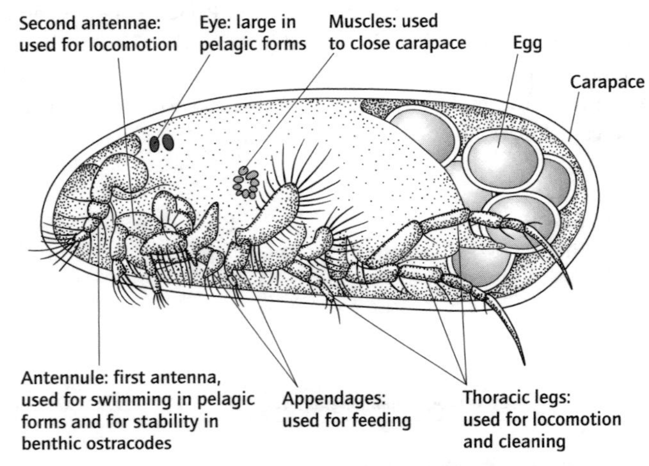

Fig. 13.9 Living ostracode morphology (female). This animal is about 1 mm in length.

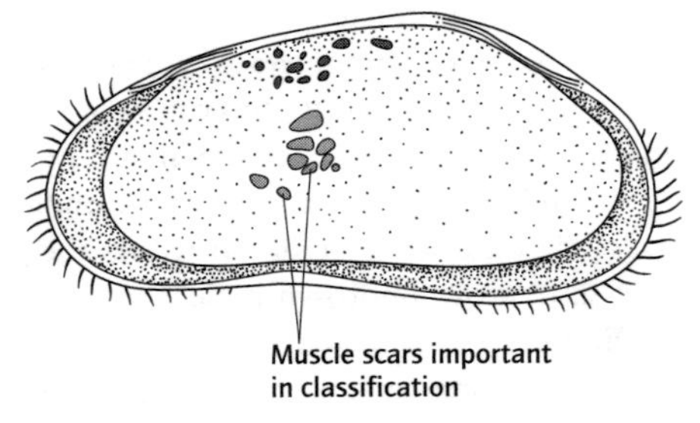

Fig. 13.10 The inner face of the valve of a female freshwater ostracode.

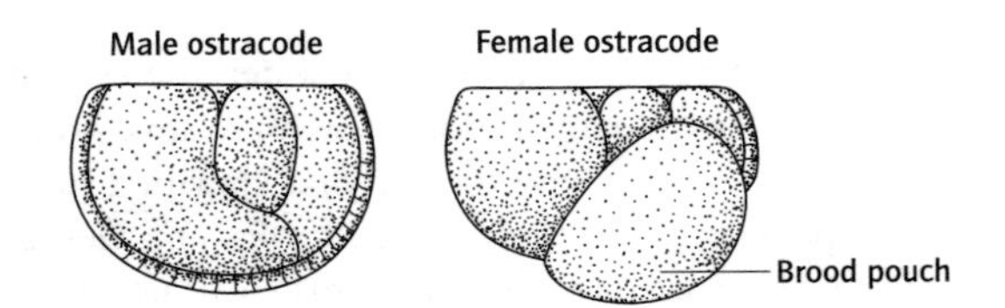

Fig. 13.11 Sexual dimorphism in ostracodes.

The soft parts are completely contained within the valves. Ostracodes are simple crustaceans with a reduced number of appendages that are mainly used for locomotion (swimming and walking) and feeding (Fig. 13.9). The valves are closed by adductor muscles, which leave scars on the interior surface (Fig. 13.10). Slender canals containing sensory bristles, or setae, perforate the carapace. Clear eye spots may also be developed on the carapace to enhance the ostracode's sensory perception. The most important sensory organs in living ostracodes are the sensory hairs that occur on the appendages and the valves. Reproductive organs occupy a large proportion of the internal volume and sexual dimorphism is common. Males usually have a greater length : height ratio. Some Palaeozoic females had a brood pouch that formed a distinctive swelling on the external surface of the carapace (Fig. 13.11). The eggs of living ostracodes are usually shed freely rather than being brooded by the female. Some freshwater species produce eggs that are very resistant to dessication. Some living ostracodes use bioluminescence to attract mates. Low flashes of bluish light are produced by external secretions. Males of the reef-dwelling genus *Vargula* are able to synchronize their flashes resulting in a dramatic display.

Ecology/palaeoecology

Ostracodes can be benthic or pelagic. Pelagic ostracodes are found only in marine environments, whereas benthic ostracodes live in marine or fresh water. Carapace structure, shape, and sculpture varies according to substrate type, salinity, temperature, and depth, making ostracodes useful indicators of palaeoenvironment.

Freshwater ostracodes tend to have thin, ovate, featureless carapaces. Marine benthic ostracodes are more heavily calcified with certain features that can be related to the nature of the substrate. Ostracodes that crawl on soft, fine-grained sediment can have a flattened ventral surface, sometimes with lateral projections to distribute their weight. Heavily calcified carapaces with ribs and spines are found in coarser substrates associated with turbulent, near-shore environments. Infaunal ostracodes typically have smooth, elongated carapaces. Pelagic ostracodes also have smooth carapaces but they tend to have a more subcircular shape. Certain species of ostracodes, with distinctive carapace morphologies, are restricted to particular salinity and depth ranges (Fig. 13.12). Living ostracodes have varied feeding habits and can be carnivores, herbivores, scavengers, or filter-feeders. Carnivorous species tend to prey on other crustaceans and snails, larger species may use their antennae to capture small fish. Herbivores mainly eat algae and filter-feeders use setae on one of the pairs of appendages to collect suspended particles.

Evolutionary history

The oldest ostracodes are reported from the early Cambrian, although the group was short lived. New forms radiated in the Ordovician continuing through the rest of the Palaeozoic. The main groups became extinct at the end of the Permian but a further radiation occurred in the Jurassic and ostracodes persist to the present day.

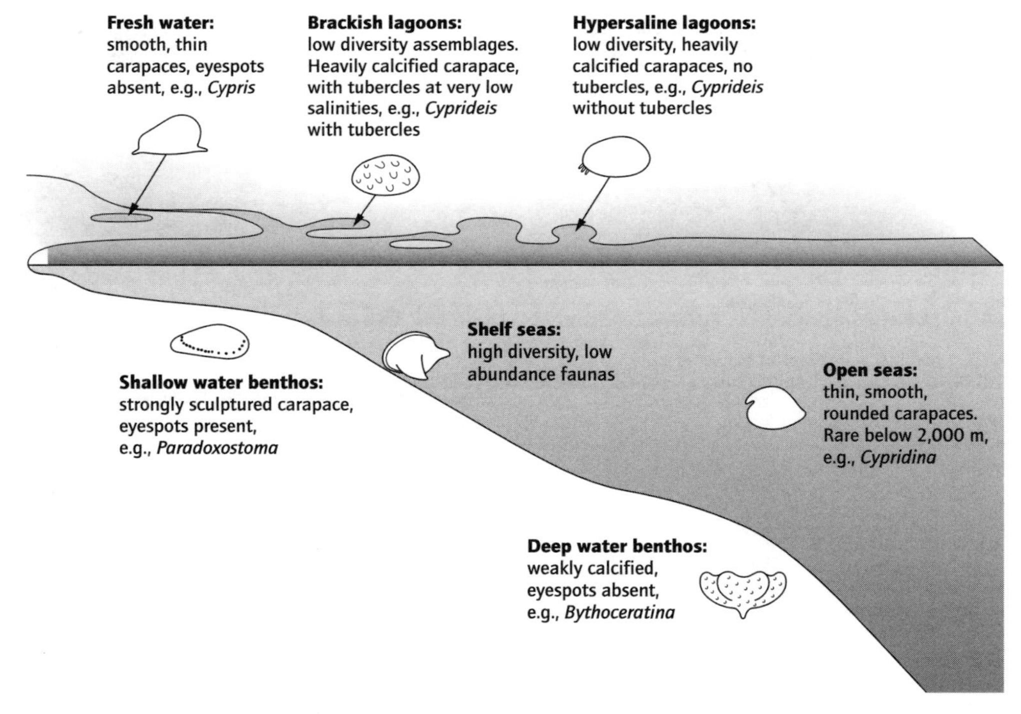

Fig. 13.12 The environmental distribution of living ostracodes.

Microvertebrates

Conodonts

Conodonts are small, phosphatic, tooth-like fossils that formed part of a complex feeding apparatus in an extinct eel-like fish. Ranging from the Cambrian to Triassic, they are important Palaeozoic biostratigraphic indicators. As their color changes with temperature of burial, they are used as indicators of thermal alteration in rocks.

Morphology

Conodonts can be divided into three basic forms (Fig. 13.13):
1 Coniform elements (cones): cusp-shaped conodonts that curve and taper to a narrow tip.
2 Ramiform elements (bars): blade-like, elongated, multi-cusped forms.
3 Pectiniform elements (platforms): broad-based, multicusped conodonts.

Rare fossil specimens show that conodont associations were arranged symmetrically. These associations, or apparatuses, were usually formed of seven or eight pairs of conodonts.

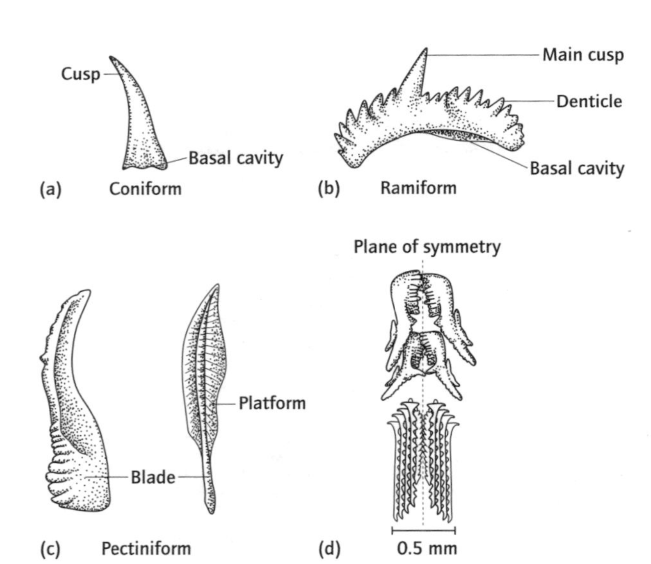

Fig. 13.13 The main forms of conodont elements: (a) *Hertzina* (6 mm); (b) *Ozarkodina* (approximately 1 mm); (c) *Polygnathus* (approximately 1 mm); and (d) natural assemblage of conodont elements in *Scottognathus typicus*.

The conodont animal

True conodont animals, preserved with the conodonts *in situ*, have been found from several localities including Scotland, UK. The animal is bilaterally symmetric with a long eel-like body, up to 55 cm in length, a small head, segmented muscles, and fins. Large eyes dominate the head and the conodont apparatus is positioned immediately behind them. The conodont animal was a vertebrate that resembled a hagfish or lamprey and the apparatus may have been used to grasp and crush prey (Fig. 13.14).

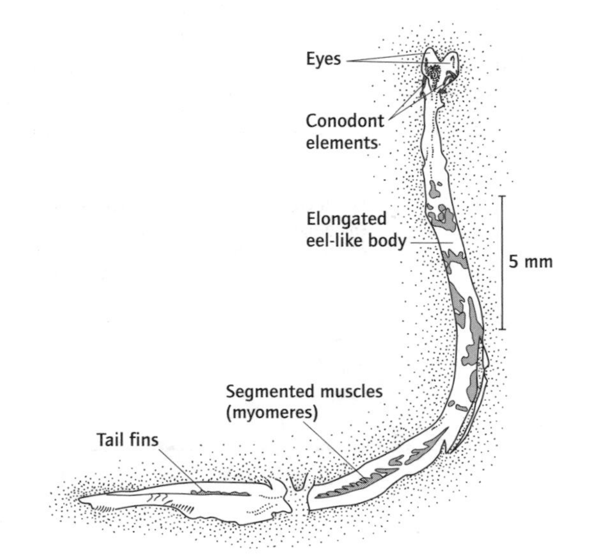

Fig. 13.14 Impression of the body of the conodont animal from the Lower Carboniferous Granton Shrimp Bed, near Edinburgh, UK.

Palaeoecology

Conodonts are found in a range of marine environments but are most common in tropical near-shore environments. Some Palaeozoic communities show a relationship with depth, and deeper water assemblages are generally less diverse.

Evolutionary history

Simple coniform elements, are known from the early Cambrian. Conodont diversity reached its maximum during the mid-Ordovician with 60 genera. Diversity declined sharply through the Silurian but conodonts radiated again during late Devonian times. After the Devonian extinction, conodont associations were less variable. Only a few forms survived the Permian mass extinction and the entire group disappeared at the end of the Triassic.

Plants

Spores and pollen

The study of spores and pollen used to reconstruct long-term vegetative changes is called palynology. Spores and pollen are part of the plant reproductive system. As they are very resistant and vast numbers are dispersed over wide areas, they are important in biostratigraphy. They can also be useful palaeoenvironmental indicators, particularly for the Quaternary.

Morphology

Spores and pollen grains are very distinctive. In general, pollen grains are smaller than spores: 25–35 μm compared with 100–200 μm in diameter. Large spores, megaspores, can be up to 4 mm in diameter. Most spores and pollens have a double-wall structure and the outer wall is very robust in order to resist desiccation and prevent microbial attack. Wall sculpture is variable. The surface can be granulated, pitted, or ornamented with rod-like projections. Apertures within the wall allow for germination of the pollen or spore and also accommodate size changes in response to variations in humidity.

Shape and aperture type form the basis of pollen and spore identification. Four main types can be identified on the basis of aperture shape (Table 13.3).

Table 13.3 Important groups of spores and pollen.

Spore/pollen	Description	Appearance
Trilete spore, e.g., ferns	Tetrahedral spores with Y-shaped aperture (trilete mark)	
Monoporate pollen, e.g., grasses	Spherical grain with single, rounded aperture	
Triporate pollen, e.g., birch	Spherical to tetrahedral grain with three, equally spaced equatorial apertures	
Saccate pollen, e.g., pines	Distinctive elongate grain with at least one spherical vesicle	

Pollen analysis

Examination of the pollen content of layered sediments, particularly from lakes and peat deposits, shows changes in the regional vegetation through time. Pollen diagrams are used to quantify this information and to record the number or percentage of grains at each selected level (Fig. 13.15). Through this method, regional and local pollen assemblage zones are established. These can then be used to recognize and date particular events within an area: for example, interglacial periods in the Quaternary and the effects of early man on the environment.

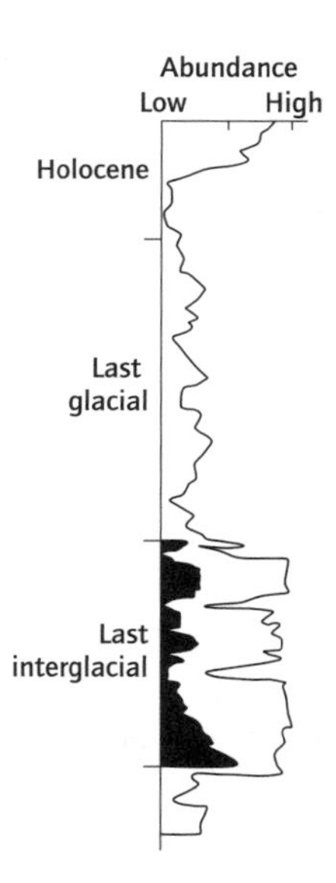

Fig. 13.15 Pollen record from Grand Pile, France. The white area represents the relative abundance of tree pollen, and the black bars show oak pollen (*Quercus*). The presence of oak pollen identifies the last interglacial period.

Evolutionary history

True spores first appeared in the early Silurian. At this time land plants were probably starting to colonize marginal marine environments. During late Silurian and early Devonian times there was an explosive radiation of land plants and diversification of spore types. Small herbaceous, seedless plants dominated the flora, some with distinctive megaspores. The first seed-bearing ferns appeared in the mid-Carboniferous and a range of pollen types was established. Classic Carboniferous swamp vegetation is known from extensive coal deposits of the period. Spores are much less abundant and pollen more common in the Permian. The decrease in spore-bearing plants has been linked with drier conditions and possible global cooling during late Palaeozoic times.

The gymnosperms dominated the Triassic and Jurassic flora. Angiosperms, flowering plants, first appeared in the early Cretaceous and gradually displaced the gymnosperms, becoming the dominant pollen producers by late Cretaceous times. The end-Cretaceous flora is similar to modern plant life although grasslands did not appear until Cenozoic times. For more information on plant evolution see Chapter 12.

Hystrichosphaeridium

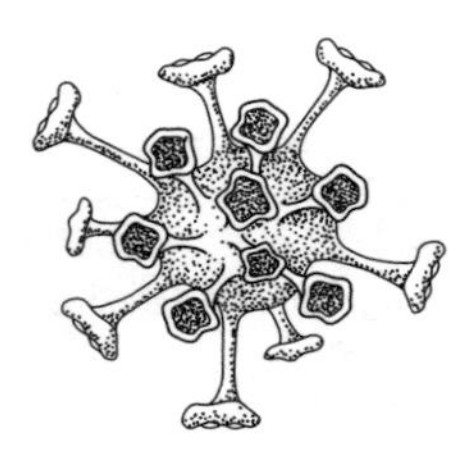

Dinoflagellate

Upper Jurassic–Middle Miocene
This fossil dinoflagellate cyst has a spherical central body (approximately 25 μm in diameter) and distinctive hollow processes that flare distally.

Coscinodiscus

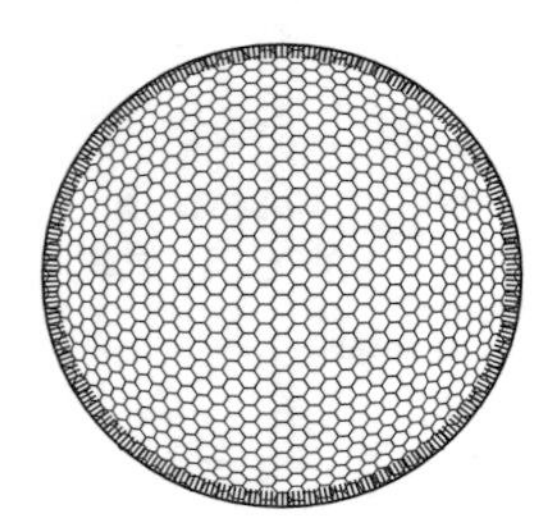

Diatom

Cretaceous–Recent
The rounded shape of this diatom shows that it belongs to the centrale group. The frustule is approximately 60 μm in diameter. This diatom is planktonic and typical of inshore and outer shelf environments.

Bathropyramis

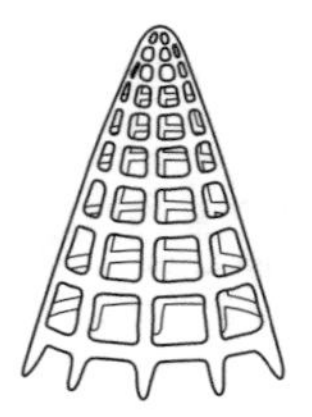

Radiolarian

Cretaceous–Recent
Bathropyramis is a radiolarian with a lattice-like skeleton, approximately 300 μm in height. Three radial spines form a tripod and the mouth is open. These cone-shaped radiolarians with an open basal mouth have adapted to live in areas with rising water currents.

Globigerina

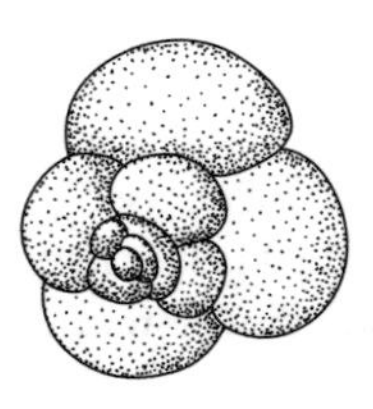

Foraminifera

Palaeocene–Recent
This thin-walled, globular foraminifera is planktonic. The test (approximately 300 μm in diameter) has been modified to maintain buoyancy. The chambers are inflated and the surface of some living species are covered with fine spines that support a frothy ectoplasm. Planktonic foraminifera are extremely important biostratigraphic markers in the Cenozoic.

Bolivina

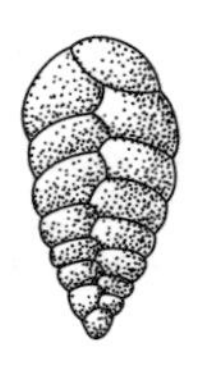

Foraminifera

Upper Cretaceous–Recent
Bolivina has an elongate, calcite test formed from two rows of chambers (length approximately 1 mm). This distinctive foraminifera is common at abyssal depths (2,500–7,000 m).

Elphidium

Foraminifera

Lower Eocene–Recent
Common in brackish water, this benthic foraminifera has a calcareous test (approximately 500 μm in diameter) with a distinctive ridged sculpture. *Elphidium* typically occurs in lagoons, estuaries, and near-shore environments.

Beyrichia

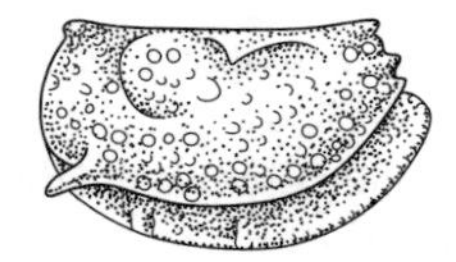

Ostracode

Lower Silurian–mid-Devonian

The carapace of this extinct marine ostracode had a long straight hinge line (about 1 mm in length) and a distinctive granular or pitted external surface. This genus shows clear sexual dimorphism. The carapace was expanded in the female to accommodate the brood pouch. It has been suggested that this ostracode swam in shallow waters and fed on detritus, benthos, and plankton.

Cypridina

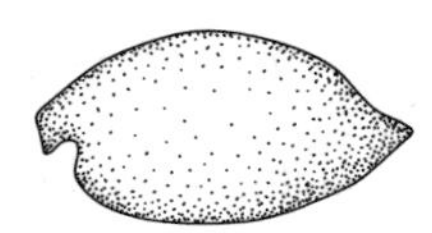

Ostracode

Upper Cretaceous–Recent

Cypridina is distinguished by a thin, featureless, ovate carapace (approximately 1.5 mm in length). This ostracode is a marine, pelagic filter-feeder, thriving in nutrient-rich waters associated with upwelling. The projection at the anterior of the carapace encloses elongated front limbs adpated for swimming. It has two stalked, compound eyes and a median simple eye.

Emiliana

Coccolithophorid

Cenozoic–Recent

The most common coccolithophore in the modern oceans. The coccosphere is about 5 μm across.

Cypris

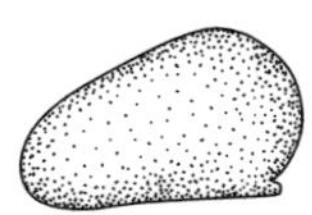

Ostracode

Jurassic?/Pleistocene–Recent

Cypris lives in freshwater environments, particularly ponds. Most freshwater ostracodes have a smooth carapace. In *Cypris* the carapace is thin, smooth, and relatively large (2.5 mm in length). The hinge mechanism is simple and eye spots are absent.

Bythocertina

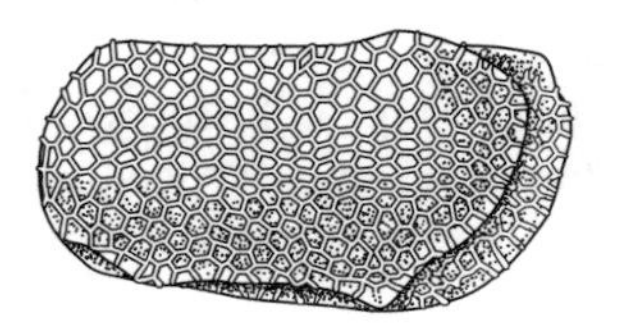

Ostracode

Upper Cretaceous–Recent

Living at depths between 2,000 and 3,000 m, these ostracodes are often referred to as psychrospheric. Such ostracodes have adapted to the lack of light, cool water conditions (4–6°C) and constant salinity. Most are blind and have large, strongly sculptured carapaces (> 1 mm). *Bythocertina* can be smooth or have a spinose or reticulate sculpture.

Quercus

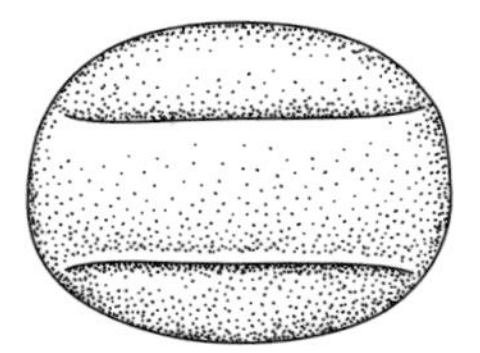

Pollen

Cenozoic–Recent

Oak pollen, typically about 12 μm across. This pollen type is indicative of warm Mediterranean-type climates.

Glossary

Antennule – front appendage in ostracodes, used for swimming or stability.
Aperture – opening in foraminifera through which most of the soft tissue projects.
Autotroph – organism that synthesizes organic compounds from inorganic materials. Such organisms are sometimes called producers.
Benthos – bottom-dwelling organisms.
Calcite compensation depth (CCD) – depth in the oceans at which calcite becomes unstable and may dissolve.
Carapace – term for "shell" used for ostracodes.
Centrales – circular diatoms with radial symmetry.
Chamber – one repeated element of calcareous foraminiferal tests.
Coniform – cusp-shaped conodont elements.
Diatomites – lithified diatom oozes.
Epifaunal – organism living on the surface of the sediment.
Frustule – term for "shell" used for diatoms.
Heterococcoliths – coccoliths formed from crystals of varying sizes.
Heterotroph – organism that consumes other organisms or decaying matter. Such organisms are sometimes called consumers.
Holococcoliths – coccoliths formed from identically sized calcite crystals.
Infaunal – organism living within sediment.
Megaspores – large spores up to 4 mm in diameter.
Nannofossils – submicroscopic fossils generally less than 50 μm.
Nannoplankton – submicroscopic floating organisms.
Nassellaria – radiolarians with conical tests.
Palaeobathymetry – depths of ancient marine deposits.
Palynology – study of spores and pollen.
Pectiniform – broad-based, multicusped conodont elements.
Pelagic – organism living within the water column.
Pennales – elliptical diatoms with bilateral symmetry.
Photic zone – lighted part of the water column.
Phytoplankton – microscopic, photosynthetic, floating organisms. The two main groups are the diatoms and dinoflagellates.
Planktonic – floating lifestyle within the water column.
Primary producers – equivalent to autotrophs. Organisms that produce organic substances from inorganic material. Primary producers are usually photosynthetic organisms.
Protoplasm – very general term for material contained within the cell.
Pseudopodia – strands of protoplasm that aid locomotion and trap food in foraminifera.
Ramiform – blade-like, multicusped conodont elements.
Setae – sensory bristles in ostracodes.
Spumellaria – radiolarians with spherical tests.
Test – term for "shell" used for foraminiferans and radiolarians.
Zone fossils – fossils that characterize particular sections of the stratigraphic record.
Zooplankton – nonphotosynthetic floating organisms, that feed on phytoplankton or suspended organic matter.

14 Trace fossils

- Trace fossils are preserved impressions of biological activity that record fossil behavior.
- Distinct trace fossil assemblages can be used in the interpretation of palaeoenvironment.
- Two main classification schemes exist: (i) behavioral (ethological); and (ii) environmental (ichnofacies).
- Traces are named according to the form of the trace rather than the tracemaker.
- Trace fossils are important in the identification of the Precambrian–Cambrian boundary.

Introduction

Trace fossils (or ichnofossils) are the biogenic sedimentary structures that record biological activities such as burrowing, walking, and feeding. As sedimentary structures, trace fossils are usually preserved *in situ* and are more accurate and reliable indicators of the sedimentary environment than body fossils, which are subject to postmortem transport. Assemblages of trace fossils form stable groupings that are widely used in the reconstruction of palaeoenvironments. However, the relationship between the fossil species and the trace fossil is complex and there is no simple correlation between species, activity, and trace. Furthermore, the name given to the trace fossil is usually not connected with the animal that made the trace. The following principles should be recognized in order to understand and apply trace fossil evidence correctly.

1 The same species can produce different traces relating to different behavior patterns.
2 The same trace may be preserved differently in different sediments.
3 Different species may produce identical traces when behaving the same.
4 Traces may be modified by changes in environmental conditions.

Trace fossil genera are called ichnogenera and species are called ichnospecies. Major groups of trace fossils are separated on the basis of general shape, orientation in the sediment, and structure of the trace. This chapter focuses on trace fossils in the marine sedimentary environment; not included here are coprolites (animal excrement), trace fossils associated with plants (for example, wood-boring and leaf-feeding), and predation marks.

Preservation

Biological traces are usually preserved at the boundary between two different types of sediment (Fig. 14.1). As traces are generally three-dimensional, the precise morphology may be difficult to ascertain, particularly if there is limited exposure or if only a cross-section of the trace exists. Entire traces preserved in three dimensions are full reliefs. They are usually burrows that have been infilled by later sediment and are preserved as casts. Collapsed burrows and traces formed on the surface of the bedding plane are semireliefs. Traces preserved on the top of the bed are epireliefs and those on the lower surface are hyporeliefs.

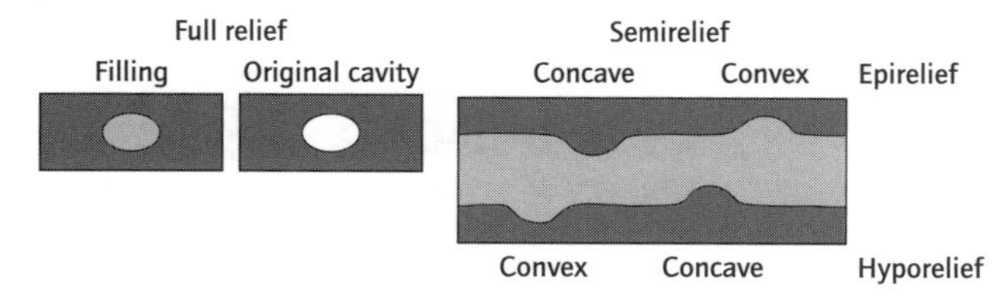

Fig. 14.1 Types of trace fossil preservation.

Ethological (behavioral) classification

This fundamental classification is founded on the behavioral characteristics represented by the trace fossils (Table 14.1). The most important categories relate to feeding, dwelling, and locomotion. As the units are divided on the basis of activity there maybe some overlap between them if the organism performed more than one behavior (for example, feeding and crawling) at the same time. Also, different parts of the trace fossil structure may fall into different categories.

Table 14.1 Ethological classification of trace fossils; 11 categories comprise the scheme. The characteristics of the eight main groups are shown. Those omitted are: praedichnia (predation traces), aedificichnia (structures built above the substrate), and calichnia (breeding nests).

Category	Description of behavior	Preservation and structure	Example ichnogera (and their possible producers)	
Cubichnia	Resting or hiding traces formed during a temporary pause in locomotion. Rather than resting, the traces usually represent feeding by stationary feeders or refuge	Usually as casts on the bottom of beds. Morphology reflects the nature of the undersurface of the resting organism	*Asteriacites* (starfish)	*Rusophycus* (trilobite)
Domichnia	Dwelling traces formed by burrowing or boring into the substrate. Traces represent the long-term home of an animal that is stationary in the burrow	Deep excavations of varying shape found within beds. Usually form an anastomosing tubular structure, or a straight or a U- shaped burrow	*Skolithos* (tube worm)	*Thalassinoides* (crustacean)
Fugichnia	Escape structures formed when animal moves rapidly upward to prevent burial by a sudden influx of sediment or, more rarely, laterally to avoid predators	Usually within beds in association with domichnia. Sediment at proximal end of burrow is reworked and lamination is disturbed	Escape structure made by bivalve	Escape structures made by polychaete
Equilibrichnia	Adjustment traces formed by infaunal animals maintaining the position of their burrow with respect to a gradually agrading or degrading sea floor	Occurs within beds in association with domichnia. Distinct laminations formed parallel with the main burrow	*Diplocraterion* Upward movement	*Diplocraterion* Downward movement
Repichnia	Locomotion traces that represent directed movement (from A to B). Includes walking, crawling, tunneling, and running	Found as casts on the bottom of beds or as epireliefs. Characteristically continuous, elongate trails, sometimes revealing motion of limbs	*Cruziana* (trilobite)	*Diplichnites* (arthropod)
Pascichnia	Grazing traces formed by systematic exploitation of a particular area of the sediment for food	Horizontal traces found on or beneath the bedding plane. Distinctive furrowed trails with sinuous, meandering form but paths rarely cross	*Nereties* (worm)	*Phycosiphon* (worm)
Agrichnia	Traps and farming traces formed by animals that trap food particles or grow algae. Structures represent feeding and dwelling in a fixed area	Complicated horizontal network of burrows with multiple openings to surface. Usually preserved as hyporeliefs on bottom bed surface	*Paleodictyon*	*Spirorhaphe*
Fodinichnia	Traces which represent two functions: deposit feeding and dwelling. Excavations are made as the organism eats the sediment and digests the food within it	Three-dimensional form usually found within the bed. Generally horizontal but more complex forms can have branched networks of burrows	*Chondrites* (worm)	*Dictyodora* (worm)

Marine ichnofacies classification

Arguably the most important use of trace fossils is their application to palaeoenvironmental interpretation. As the behavior of an organism is linked to certain environmental conditions, so the manifestation of this behavior (i.e., the trace fossil) is indicative of particular environments. Different types of trace fossils form distinct groupings that are generally stable through time. Four of the identified ichnofacies are primarily related to energy conditions, water depth, and deposition rates. Other ichnofacies reflect the nature of the substrate. Each ichnofacies is named after a characteristic ichnogenus (Fig. 14.2).

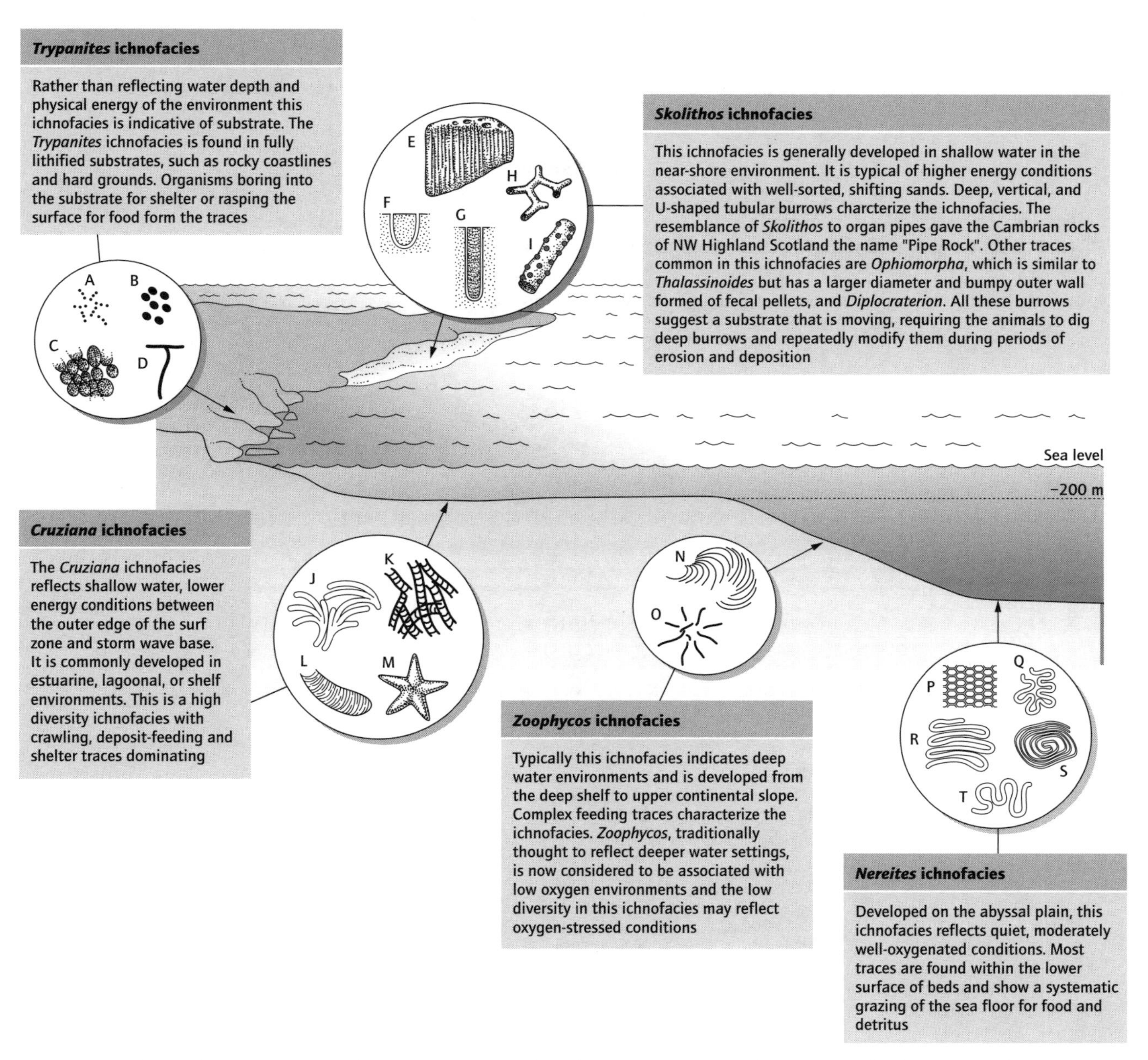

Fig. 14.2 Ichnofacies and their main trace fossils: A, *Caulostrepsis*; B, echinoid borings; C, *Entobia*; D, *Trypanites*; E, *Skolithos*; F, *Arenicolites*; G, *Diplocraterion*; H, *Thalassinoides*; I, *Ophiomorpha*; J, *Phycodes*; K, *Crossopodia*; L, *Rhizocorallium*; M, *Asteriacites*; N, *Zoophycos*; O, *Lorenzinia*; P, *Paleodictyon*; Q, *Cosmorhaphe*; R, *Helminthoida*; S, *Spirorhaphe*; T, *Taphrhelminthopsis*.

Evolution of trace fossils

As trace fossils are controlled by the nature of the depositional environment, they do not make good zone fossils. The only exception to this is at the Precambrian–Cambrian boundary. The base of the Cambrian is defined by the first appearance of the feeding trace of a worm-like animal, *Treptichnus pedum*. There is a significant increase in the diversity and complexity of trace fossil assemblages found in basal Cambrian rocks contrasting with the simple burrows typical of late Precambrian times. Although trace fossil assemblages have remained generally stable through time there are three major trends which can be identified in the Phanerozoic. Firstly, there is an increase in diversity at the end of the Ordovician associated with the radiation of metazoans. Secondly, the depth of infaunal burrowing increases in the early Carboniferous, with deep burrowers appearing in the late Palaeozoic. And, thirdly, there is a tendency for onshore traces to move into the offshore environment. For example, *Zoophycos*, which is typical of the inner shelf environment in the Ordovican had moved to deep slope environments by the Tertiary, and *Paleodictyon* migrated from the shallow shelf into deeper water early in the Cambrian.

Glossary

Epirelief – trace preserved on the top of the bed.
Full relief – entire trace preserved in three dimensions within the sediment.
Hyporelief – trace preserved on the lower surface of the bed.
Ichnogenera – trace fossil genera.
Ichnospecies – trace fossil species.
Semirelief – collapsed burrow or trace formed on the bed surface.

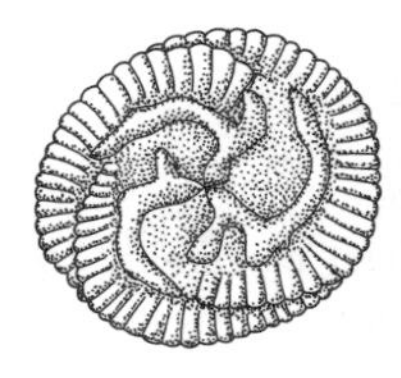

15 Precambrian life

- Life evolved early in the history of the Earth, possibly before 3.8 billion years ago.
- Complicated, eukaryotic cells appear in the geological record 2.7 billion years ago.
- The time of origin of multicellular animals is in dispute, with different estimates being derived from molecular and palaeontological techniques.
- The evolution of photosynthesis had the profound effect of adding oxygen to the Precambrian atmosphere.

Introduction

The Precambrian comprises most of the history of the Earth. During this time life evolved, diversified, and began to change the atmosphere and surface of the planet.

The Earth formed 4.55 billion years ago and for the succeeding 650 million years it was bombarded by large meteorites. These would have been of sufficient magnitude to have vaporized liquid water on the surface, and possibly to have melted the solid crust. After a final dramatic bombardment, large meteorites ceased to hit the Earth around 3.9 billion years ago. The oldest sedimentary rocks preserved on the Earth's surface have an age of about 3.8 billion years. These sediments show that there was liquid water on the surface and an atmosphere that lacked oxygen.

Modern organisms can point, to some degree, to what early life might have been like. Conservative gene sequences, which are used to identify the earliest ancestors of living organisms, suggest a single common ancestor to all modern life forms. The closest modern relatives to this common ancestor are prokaryotes, simple single-celled organisms inside which there is little internal organization. In addition, these simple modern bacteria usually live where oxygen is absent and can be tolerant of very high temperatures, in some cases approaching 100°C. Perhaps we should, therefore, be looking for early life in the form of simple, small cells in areas characterized by high heat flow, such as around volcanic hot springs or deep sea hydrothermal vents.

Although fossils are sparse in most Precambrian rocks, evidence for the activity of life on Earth is abundant. There is unambiguous evidence for the build-up of oxygen in the Precambrian atmosphere, though the timing and timescale of the process are disputed. The oxygen was produced by photosynthetic, single-celled organisms. Huge deposits of oxidized iron, known as banded iron formations, formed between 3.5 and 1.8 billion years ago and there may have been a biological source for the oxygen they consumed. As these deposits declined in abundance with time, so did the presence of reduced iron in sedimentary rocks, showing that free oxygen was beginning to remain in the atmosphere.

As atmospheric oxygen began to accumulate, it facilitated the radiation of complicated cellular life – eukaryotes, and eventually multicelled animals and plants. These organisms need oxygen for efficient respiration. The oldest known eukaryotes left distinctive chemical fossils in rocks that are 2.7 billion years old in Australia.

The fossil record of multicellular organisms is controversial. A filamentous fossil called *Grypania* is probably the remains of a multicellular plant. It is found in rocks up to 1.4 billion years old. The best known multicellular fossils of the Precambrian are much younger, found in rocks dated at betweeen 590 and 543 Ma. These are the Ediacaran faunas, a group of enigmatic organisms with a worldwide distribution and unique body forms.

Animals with mineralized skeletons, like most of those we see around us today, are almost absent from the Precambrian rock record, but the appearance of diverse forms of life with hard parts at the Precambrian–Cambrian boundary shows that animals had probably been diversifying for some time without leaving traces in the rocks.

Evidence for early life

The search for life's origins, and evidence for the pathways of its early evolution are two of the most important fields in palaeontology, especially as this research informs the search for life on other planets. Evidence from a variety of sources is used to develop the debate, but as yet no consensus has emerged.

Molecular evidence points to a single common ancestor for all life on Earth, which makes it extremely likely that life evolved here, rather than being brought from elsewhere, as has been suggested in the past. This, in turn, constrains the timing of the origin of life on Earth. From its formation until around 3.9 billion years ago the Earth was bombarded by enormous meteorites. It is considered unlikely that life could have survived this bombardment, so life must have evolved less than 3.9 billion years ago. Unusual carbon spheres from 3.8 billion-year-old rocks at Isua in Greenland are interpreted by some workers as the remains of simple cells. Cell chains and stromatolites have been reported from rocks around 3.5 billion years old from Australia (Fig. 15.1), though these now appear to be artifacts.

Isotopic evidence suggests that life evolved the ability to produce complicated biochemical pathways early in its history. Modern photosynthesizing organisms draw CO_2 from the atmosphere, and preferentially extract more of the lighter stable isotope of carbon, ^{12}C. A light carbon isotopic signature is evidence for photosynthesizing organisms. It may be present in the Isua rocks, and is certainly common in rocks less than 3.7 billion years old. The importance of photosynthesis is that a common by-product is oxygen.

Biological evidence cannot accurately date the origin of life, but suggests that it lies within a group of simple bacteria, or prokaryotes (Fig. 15.2), called hyperthermophiles, which are adapted to life at high temperatures. This may indicate that the location of the origin of life was in hot spring environments, or close to mid-ocean ridges.

Taking these lines of evidence together suggests that life evolved soon after Earth became stable enough to sustain it. They also suggest that life quickly passed through a range of simple stages to produce complicated metabolisms within 200 million years of its origin.

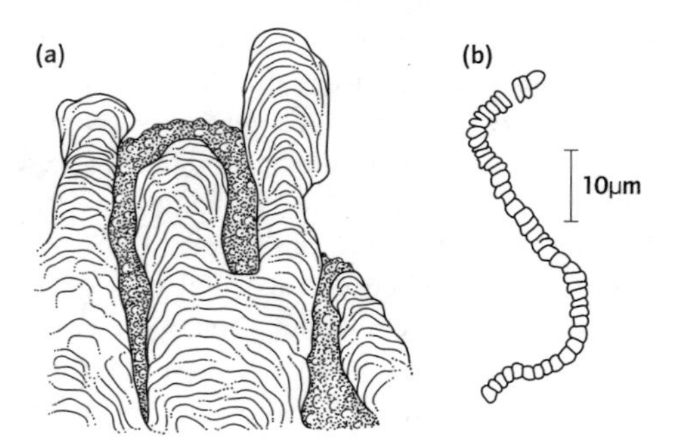

Fig. 15.1 Possible early fossils. (a) An Archean stromatolite, and (b) a string of what may be prokaryote cells.

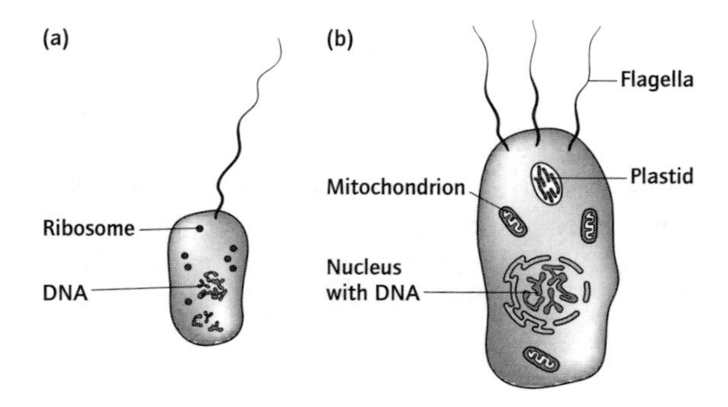

Fig. 15.2 (a) A prokaryote, and (b) a eukaryote. Note the internal organization and larger size of the eukaryote.

Table 15.1 Summary of the main lines of evidence for the timing of the evolution of life.

Formation of the Earth	The Earth formed 4.6 Ga ago, and it is most likely that life evolved after this point. In addition, the intermittant collision of huge meteorites with the Earth continued until about 3.9 Ga. This evidence suggests that life evolved more recently than this
Geological	The oldest rocks that give evidence of surface conditions are the Isua Supracrustals from Greenland. These show that by 3.8 Ba there was liquid water at the Earth's surface and a largely reducing atmosphere. In addition they contain unusual carbon spheres and an isotopic signature, which might be evidence for the existence of life, though this is disputed by some workers
Isotopic	Most living organisms fractionate isotopes of the elements they extract from the environment. In particular, the enzyme responsible for extracting carbon dioxide from air during photosynthesis preferentially extracts the lighter isotope of carbon. This is preserved when the plant remains are buried, in carbon or carbonates. A light C isotope signature is good evidence for the existence of life and is common in rocks younger than 3.7 Ba
Biological	The comparison of parts of the genetic code shared by all organisms can give a possible date for the time when the last common ancestor of modern organisms lived. Unfortunately estimates are currently widely divergent, giving dates between 1 and 6 Ga

The origin of complexity

Tiny prokaryotes are still the most numerous life form on Earth. The generally larger eukaryotes, as single cells or larger multicellular organisms, evolved later. The origin of eukaryotic life is a source of intense debate.

Eukaryotes contain organelles that perform specific functions. Energy storage is undertaken by mitochondria, photosynthesis by chloroplasts, movement by cilia or a long flagellum, and reproduction and the storage of genetic information within a nucleus (Fig. 15.2). The presence of a nucleus is one of the defining features of this group of organisms. Some eukaryotes can combine their DNA with that of another individual, that is they can reproduce sexually, increasing the potential genetic variability. In addition, the cells of eukaryotes contain a cytoskeleton, made up of tiny tubes and proteins, which allows the cell to dispense with a rigid outer wall. As a result, eukaryotes can change their shape, even to the extent of surrounding and consuming other cells.

The origin of eukaryotes was almost certainly by a process known as endosymbiosis. Individual prokaryotic organisms were incorporated into eukaryotic cells to produce an interdependent entity whose functions were carried out by specialized organelles derived from the absorbed organisms (Fig. 15.3). The mitochondria and chloroplasts of eukaryotic cells have an independent genetic code reflecting their origins as independent organisms. In addition, some modern prokaryotes bear a marked resemblance to the organelles of eukaryotic cells.

Most single-celled eukaryotes are relatively large, usually between 10 and 100 μm in size. As the eukaryotic cell size increases, so its surface area to volume relationship becomes less favorable so that the movement of molecules across the cell membrane by diffusion will not take place quickly enough to supply the cell. Instead, active transport mechanisms are required to import material from outside the cell. The increase in the cell's energy demands that follows from increasing size is satisfied by the development of oxidative energy production. The citric acid cycle (Kreb's cycle) produces far more energy than the anaerobic, glycolitic cycle used by primitive prokaryotes. Eukaryotic diversification could only have occurred after the oceans and the atmosphere had became oxygenated.

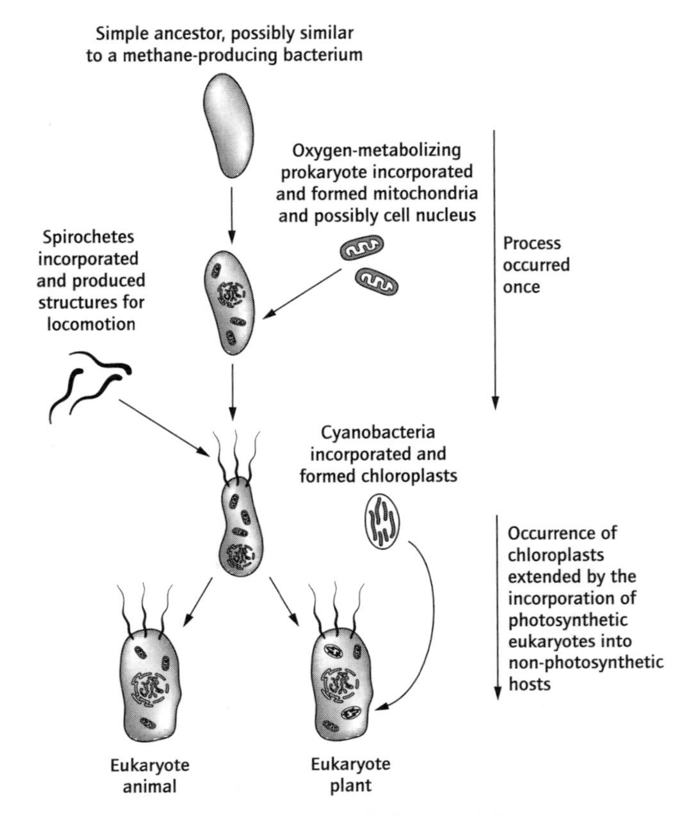

Fig. 15.3 The endosymbiosis theory of eukaryote origins.

It is likely that gaseous oxygen began to accumulate in the Earth's atmosphere around 2 billion years ago. This might be the point where eukaryotes began to be able to spread geographically and to evolve rapidly as they encountered new environments.

The oldest evidence for eukaryotes comes from chemical fossils 2.7 billion years old. The oldest body fossils are acritarchs. These are thought to be the resting stages of eukaryotic planktonic algae, similar to modern dinoflagellates. The oldest acritarchs are 2.5 billion years old, and they appear to have undergone a major radiation at about 1 billion years ago, becoming common in the fossil record at this time.

Multicellular animals

Multicellular plants and animals are composed of eukaryotic cells. Their origin must therefore postdate the origin of this cell type. It is probable that all living multicellular animals shared a single common ancestor. Molecular data provide widely divergent results, but suggest that this common ancestor arose between 1 and 1.6 billion years ago.

The fossil record of metazoans is more recent, and older fossils are sometimes regarded with suspicion by experts. The oldest common and widely accepted metazoans are members of the Ediacaran fauna (see opposite), found in rocks dated at between 565 and 543 million years old.

Classifying animals

A family tree for organisms (implied in most systems of classification) provides a series of events that must have happened in the past in a known order. Classification of the metazoa is therefore a key element in deducing information about their origin and early history. For example, it is now known that animals with a gut evolved relatively late in the history of metazoans. This information makes the discovery of possible fecal pellets produced from a gut-bearing animal in rocks up to 1.9 billion years old an extremely important one, though, as usual, it is the subject of intense dispute.

Recently, large strides have been made in producing a generally agreed classification for all animals. This points to various early events in their evolution and the order in which they occurred (Fig. 15.4). Dates derived from molecular clocks can be added to this classification to suggest the timing of the important events implied by it. The main problem with this classification is the place in it filled by Ediacaran organisms. Whilst their location within such a classification is unclear, the geological evidence they offer about the origin of animals remains difficult to interpret.

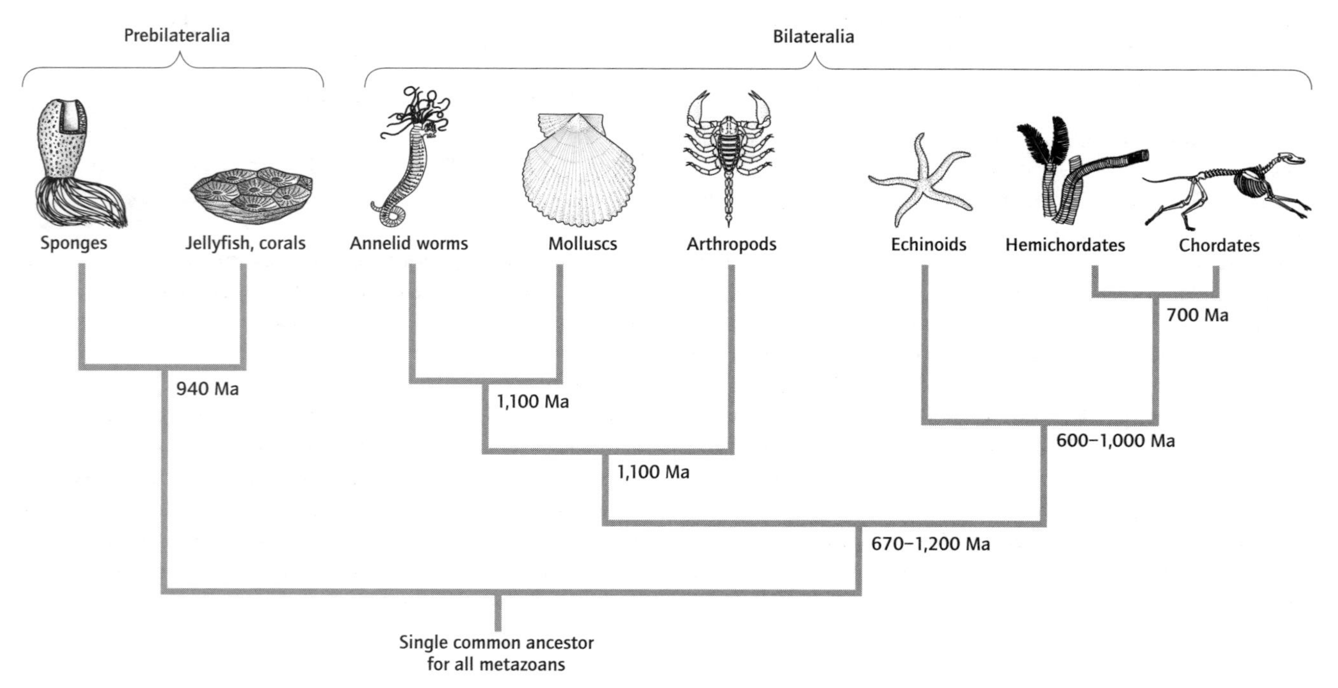

Fig. 15.4 A recent classification of multicellular animals. This classification divides animals into two broad groups – those with bilateral symmetry and those with radial symmetry. There is a large genetic gap between the two groups. The position of Ediacaran organisms is uncertain. Dates added to the diagram represent molecular estimates of time since divergence. Note the wide variation in these estimates.

Ediacaran fauna

The Ediacaran fauna represents the oldest, well known, diverse set of multicellular organisms. They are found almost worldwide, distributed in rocks from just above the tillites of the last Precambrian glaciation to the Precambrian–Cambrian boundary, and possibly beyond. Most Ediacaran assemblages are probably 565–543 million years old.

The fauna is entirely soft bodied and contains up to 30 species of organism. The lack of hard parts means that the animals were probably fossilized close to where they lived, as the dead bodies would have been too frail to have been transported for any significant time or distance. If this is, indeed, the case then the fossil assemblages represent ecosystems, rather than chance accumulations of organisms.

Two divergent views are held about the classification of these organisms. At one extreme, it is suggested that they represent early examples of modern groups, especially soft corals and jellyfish. At the other extreme, it is suggested that they comprise a kingdom of their own, called the Vendobionta, and bear little relationship to modern metazoans.

Structurally, three groups of organisms seem to be present in the fauna: (i) those with radial symmetry, which may be related to jellyfish or corals; (ii) those with bilateral symmetry, possibly related to more advanced organisms such as worms or arthropods; and (iii) those with an unusual symmetry that may not be represented by living groups. The significance of this classification based on symmetry is that modern high-level classification relies partly on this feature (Fig. 15.4). Thus jellyfish and their relatives are regarded as having a long, separate evolutionary history to other metazoans because they have a different type of symmetry in their body plan. The same principles can be applied to Ediacaran animals. If three different types of symmetry really are present in these assemblages, they must represent the end products of a considerable evolution of multicelled animals.

An example of the radial group of species is *Cyclomedusa*, an organism with concentric rings, possibly a jellyfish. An example of the bilateral group is *Spriggina*. *Tribrachidium* shows an unusual, in this case threefold, symmetry. *Dickinsonia* is an organism that can be interpreted as either a member of a modern group or something totally different (Fig. 15.5). It superficially resembles a polychaete worm, although it may be more similar to a coral polyp. However, close examination of its segments suggests that they interfingered with one another and may have had a support function, like the stitching of a quilt. This condition is seen in no living organism. Profoundly unusual symmetries such as this may imply that some of the Ediacaran organisms should be classified in a major group of their own, separate from all living metazoans.

The mode of life and ecology of Ediacaran animals is as problematic as their classification. Most of the organisms were probably benthic, but a few were planktonic. It is suggested that all of the Ediacarans lacked a gut, and that they lived a relatively quiet life, absorbing food and undergoing gaseous exchange directly across their body surfaces. Simple flatworms live like this today. However, it is also suggested that the quilted segments of some Ediacaran animals were used as algal farms, or even that some of the organisms were lichen and not really animals at all.

The Ediacaran animals reached their maximum diversity just before the Precambrian–Cambrian boundary. In the early Cambrian they were superseded by newly skeletonized, recognizably modern, organisms, that came to dominate the seas. Possible Ediacaran survivors into the Middle Cambrian form a minor component of these more diverse modern faunas.

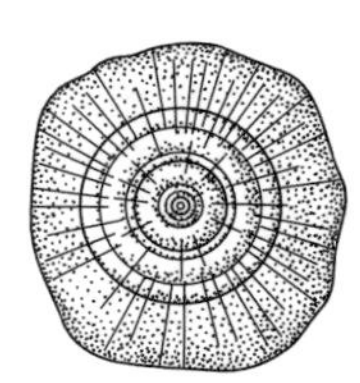

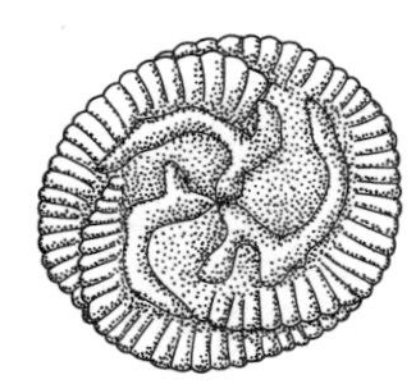

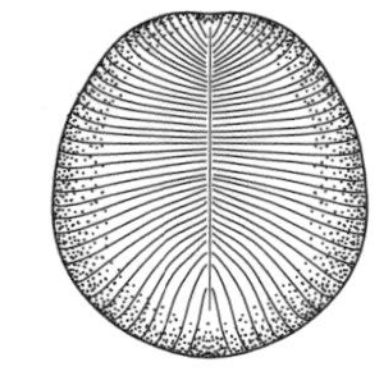

Fig. 15.5 Sketches of some representative members of the Ediacaran fauna.

Glossary

Bilateralia – organisms that display bilateral symmetry.
DNA (deoxyribonucleic acid) – double-stranded nucleic acid that can replicate itself. It determines the inherited structure of a cell's proteins. RNA is a single-stranded, simpler alternative which also carries genetic information.
Enzyme – organic catalyst that changes the rate of metabolism.
Eukaryote – organism made up of one or more cells with a nucleus enclosed by a membrane and membrane-bounded organelles.
Gene sequences – strings of DNA or RNA, which contain genetic information.
Genetic code – sequence of triplets of molecules carried on chromosomes, which determines most of the physical characteristics of an organism, and is composed of DNA or RNA.
Isotopes – elements with similar chemical properties but different numbers of neutrons.
Kreb's cycle – chemical cycle that completes the breakdown of glucose to carbon. It takes place in the eukaryotic mitochondria.
Metazoans – multicellular eukaryotes that rely on an external source of food.
Mitochondria – eukaryotic organelles that serve as the site for the operation of Kreb's cycle.
Molecular clock – method of comparing DNA sequences in order to establish the timing of evolutionary divergence.
Organelle – one of several specialized structures found in eukaryotic cells, for example mitochondria and chloroplasts.
Photosynthesis – biological process that converts light energy into chemical energy, which is stored in glucose or other organic compounds.
Plastids – eukaryotic plant organelles including chloroplasts.
Prebilateralia – animals that display radial symmetry.
Prokaryote – cell without a nucleus and membrane-bound organelles.

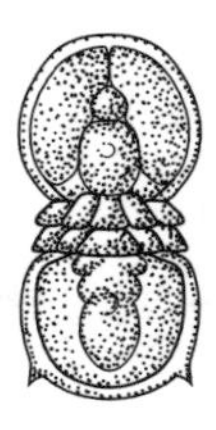

16 Phanerozoic life

- The appearance of skeletons marks the beginning of the Phanerozoic, a word meaning "revealed life".
- Subsequently, life in the oceans diversified in three major steps, each characterized by a separate fauna – the Cambrian fauna, the Palaeozoic fauna, and the Modern fauna.
- Overall diversity has increased through the Phanerozoic, punctuated by intermittent, brief, severe declines in diversity caused by mass extinctions.
- The most important innovation of the Phanerozoic was the migration of organisms on to land during the Palaeozoic.

Introduction

Close to the base of the Cambrian, many different forms of animal evolved mineralized skeletons. Hard tissues can function for attack or defense. They allow organisms to move faster and grow to a larger size than would otherwise be possible. Within a few million years, organisms were living in a wide range of diverse niches throughout the oceans of the world.

Within the oceans, life has diversified since this point in three main bursts. A Cambrian fauna, characterized by a mud-grubbing lifestyle, was replaced by a Palaeozoic fauna with a much larger filter-feeding cohort and more predators. This in turn was replaced by a Modern fauna characterized by an abundance of predators and an increasing tendency to burrow for protection.

The Cambrian fauna and the Palaeozoic fauna each seem to reach a plateau of diversity, while diversity within the Modern fauna appears to have increased since its appearance in the early Triassic. Although this is likely to be a real phenomenon, it is important to be aware of the danger of artifacts in the fossil record, which can make interpretation difficult. The greater proportion of recognizably modern species and the generally greater quantity of rock available from which to sample, can both contribute to making younger fossil assemblages appear more diverse than older ones.

Intermittently, there have been major, rapid declines in diversity — mass extinctions. These wiped out a large number of successful species and allowed a new cohort of organisms to radiate. These extinction events appear to differ in nature as well as in extent from background extinction. Characteristics that favor survival at normal times, such as specialization to a local environment, make an animal more prone to extinction during a large extinction event. While most background extinction appears to be the result of competition between species, mass extinctions are brought about through abiotic processes such as volcanic eruptions or a meteorite impact. These large events are more than scaled-up background extinctions. They involve the large-scale collapse of ecosystems.

Species diversity returns to its previous level within a short time, usually within 10 million years of the mass extinction. This kind of rapid evolution may also be different in some respects from that which occurs normally, as organisms have a chance to evolve to fill empty ecological niches rather than encountering competition.

The Cambrian explosion

More and more data are helping to explain the event at the base of the Cambrian where mineralized skeletons evolved. A few sections worldwide record the full timespan from just before to just after the Precambrian–Cambrian boundary. As these are investigated, a detailed picture of events associated with this time period is emerging.

The two most important questions about the boundary are:

1 What caused the appearance of skeletons? Was it a biological phenomenon or one related to changes in the oceans or atmosphere?

2 Did modern groups of multicellular animals exist and were

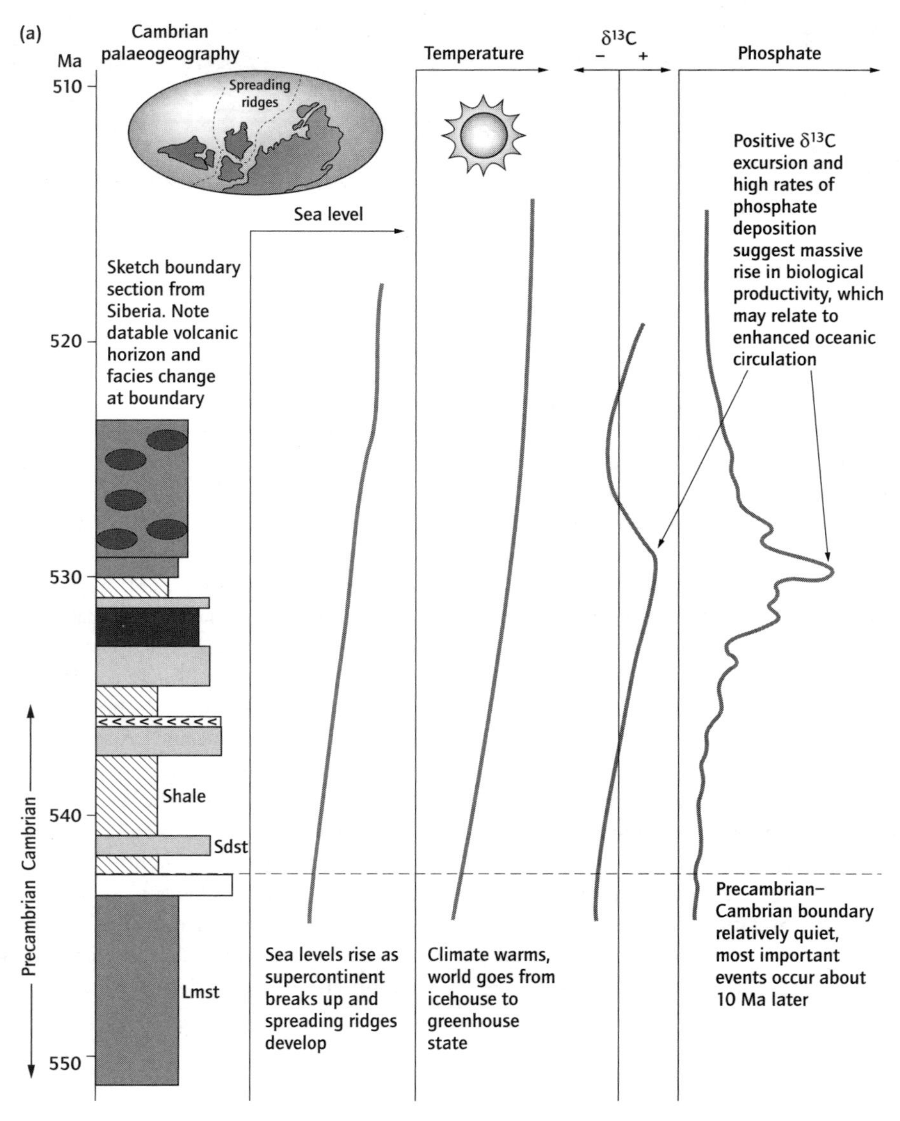

Fig. 16.1 Schematic diagram showing important elements of (a) the physical, and (b) the biological world across the Precambrian–Cambrian boundary.

they diverse prior to the boundary, or does this event signify not just the appearance of skeletons but also the appearance of the modern cohort of animal phyla?

The answer to the second question is that there is good evidence for a period of diversification of metazoans during the Precambrian. All molecular clocks suggest a Precambrian origin for the group, and many different phyla evolved mineralized skeletons in a short period of time. In addition, early Cambrian faunas are provincial, with species being specific to particular geographic regions. Evolution must have occurred before mineralization for this provincialism to have developed.

The answer to the first question is more problematic, and has yet to be resolved. The main features of the Precambrian–Cambrian boundary are shown in Fig. 16.1.

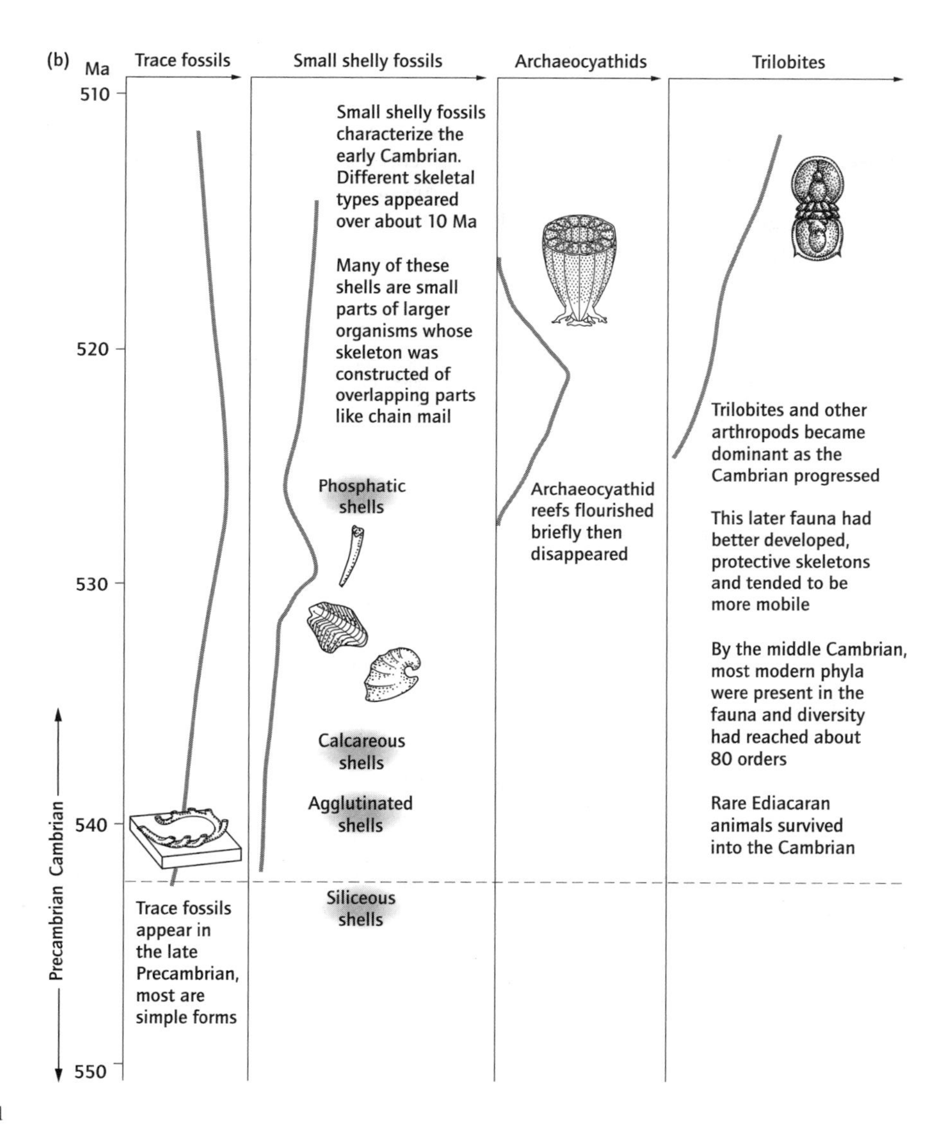

Fig. 16.1 Continued

Phanerozoic diversity

The evolution of hard parts greatly improved the fossil record. The diversity patterns of the Phanerozoic seas were documented by Professor Jack Sepkoski, and the resulting diagram is usually referred to as Sepkoski's curves (Fig. 16.2). Diversity is the net product of the originations of species minus extinctions at any one time.

This graph shows that species diversity has increased overall through the Phanerozoic. Over 800 families are now known, as opposed to around 200 in the Cambrian. However, this increase in diversity has not been uniform. There have been at least two periods of apparent stasis, one during the Cambrian and early Ordovician, and one through the Silurian to Permian, when diversity seems to have reached a temporary maximum. In addition, there have been major falls in species diversity as a result of mass extinctions. The largest of these was the end-Permian event, when species-level diversity dropped by around 95%, and family-level diversity fell by more than half.

Abiotic effects seem to have influenced the shape of these diversity curves. An obvious example is the apparent relationship between the formation of the supercontinent of Pangea and the end-Permian mass extinction. Major continental breakups, such as those of Pangea through the early Mesozoic, and of Rodinia during the Cambrian, seem to have resulted in increased adaptive radiation. This may be because continents were becoming isolated, causing biotas to evolve independently and increasing species diversity. Conversely, as supercontinents form the range of habitats declines and species diversity decreases. Changes from icehouse to greenhouse states may also have had an impact on species diversity, though this is a less clear relationship. The abrupt transition from greenhouse to icehouse in the late Ordovician may have brought about the end-Ordovician mass extinction. However, other similar transitions, for example in the Carboniferous, appear to have had no effect.

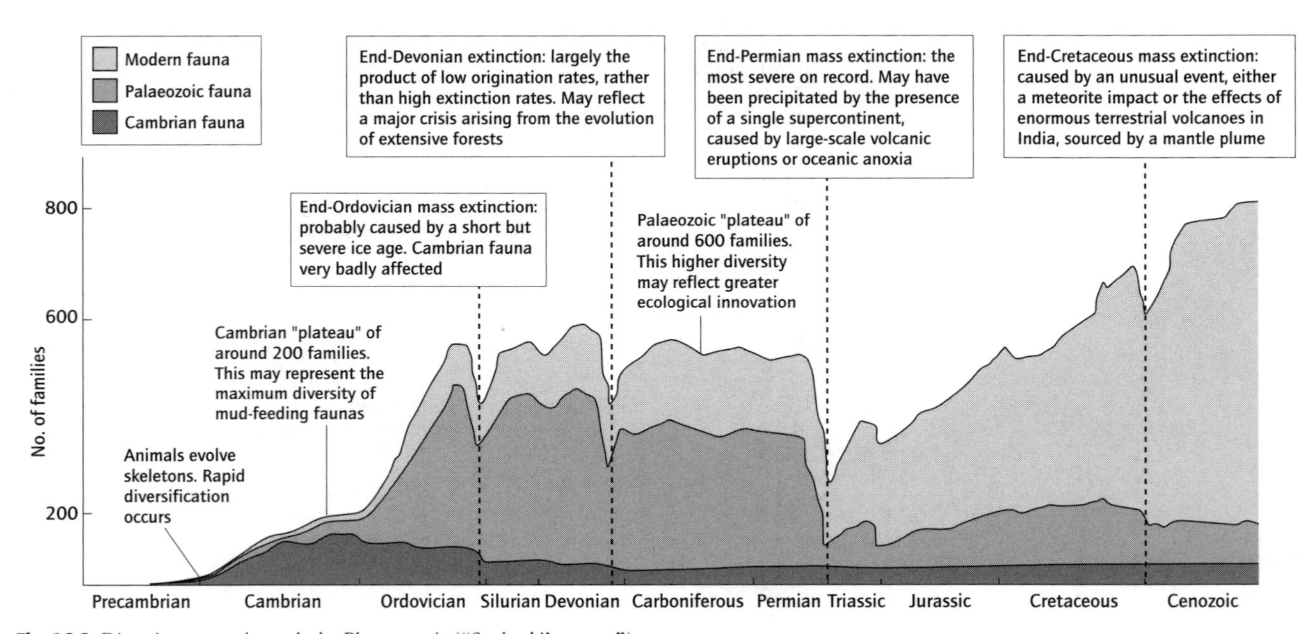

Fig. 16.2 Diversity curves through the Phanerozoic ("Sepkoski's curves").

(a) **Cambrian fauna**

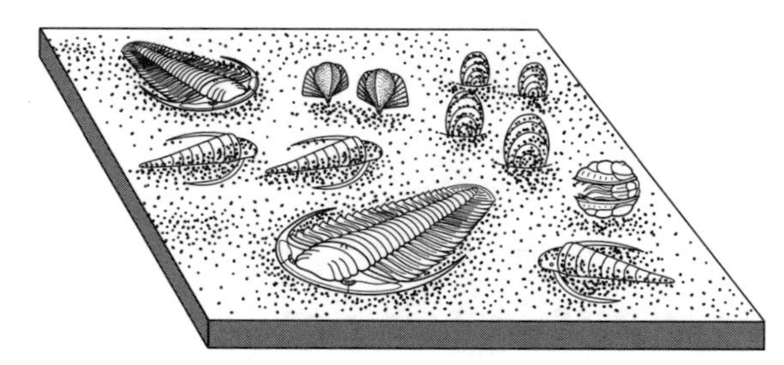

Cambrian fauna dominated by trilobites and other arthropods. Also inarticulate brachiopods and hyolithids. Fauna is predominantly based on mud eating and most animals live on the sediment surface. However, a much higher diversity of animals and niches is recorded by sites of exceptional preservation, such as the Burgess Shale, which may mean that this picture is oversimplified

(b) **Palaeozoic fauna**

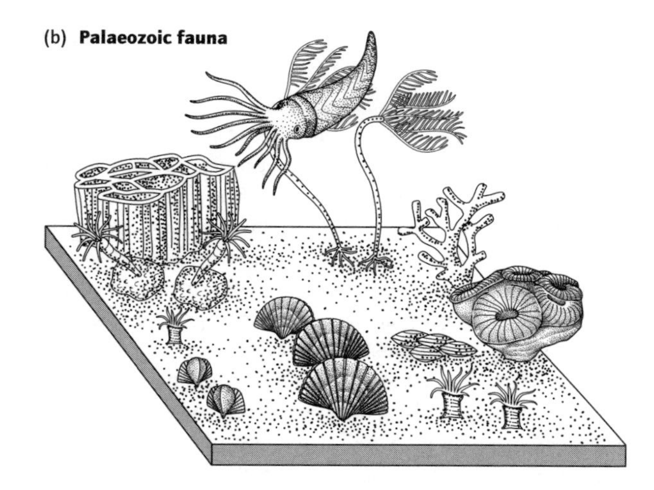

Palaeozoic fauna dominated by corals, crinoids, brachiopods, predatory cephalopods, and by graptolites in deeper water. This fauna is more tiered than that of the Cambrian, with organisms living on, and growing above, the sea bed. Reefs and other biotherms become common. Many organisms filter-feed from seawater, whilst others forage on the sediment surface

(c) **Modern fauna**

Modern fauna dominated by bivalves, echinoids, gastropods, cephalopods, and vertebrates. Burrowing and predation become very common, increasing faunal tiering yet again and keeping diversity high by regulating the abundance of prey species

Fig. 16.3 Sketches of (a) Cambrian, (b) Palaeozoic, and (c) Modern evolutionary faunas.

Biological events have also been extremely important in controlling diversity. Three different faunas can be identified through the Phanerozoic (Fig. 16.3). These faunas share common patterns of evolution and extinction and may have constituted real evolutionary assemblages of organisms living in the shallow seas. The Cambrian fauna was the first to appear and was dominated by arthropods, particularly trilobites. Most of its common animals were mud-grubbers, living on the sea floor. This fauna was replaced during the Ordovician period by the Palaeozoic fauna, which included a greater proportion of filter-feeders and predators. In turn, this fauna was replaced after the end-Permian extinction by the Modern fauna, characterized by aggressive predators and burrowing invertebrates.

Each of these three faunas exploited an increased number of potential niches available in the shallow shelves. They did this physically, by growing above and below the sea floor, and biologically, by introducing innovations of feeding method and increasing predation. These innovations offer a good explanation of the three-phase increase in diversity seen in the overall diversity curve.

Diversification

Diversity is a measure of variety. It is usually assessed at a taxonomic level, for example, the number of species or families living at a particular time. It could also be applied to individual variation, or even to variations in DNA, but variation at this level is not generally used in the interpretation of the fossil record.

All organisms share a single common ancestor that probably lived around 3.8 billion years ago. At this stage in evolution, diversity would have been extremely low by any measure. Its increase since that time must ultimately be a function of evolution, which is the means by which new species appear.

How does diversity increase? It may be that this is always accomplished by Darwinian selection. In this model, selection pressures acting on individuals result in changes in the phenotype. If these changes occur in a single lineage of animals, then there is no net change in diversity, but if the lineage splits then diversity is increased. It is believed that most lineage splits occur due to the physical separation of populations from the rest of the gene pool. This may be as a result of migration, or as a result of the erection of physical barriers by geological processes.

The breakup of a major continent into smaller elements is an obvious way to physically separate populations of the same species. It is notable that such continental breakups do seem to relate to episodes of diversification (see p. 140). However, such separations must often happen at smaller scales, for instance when falling water levels split a single lake into two, or when changes in drainage pattern separate two river systems.

There is also an important biological instrument for increasing diversity. This is the origin of new potential, either in the form of new anatomical characteristics, or in the form of new modes of behavior. The origin of a burrowing habit in members of the Modern fauna, or the acquisition of flight by birds, or sight in trilobites, for example, all led to increases in diversity.

The rate of diversity change appears to be linked to the causes of change, and even more clearly to the degree of competition experienced by organisms. Broadly, the most rapid diversification occurs when ecospace is empty, and slower diversification occurs when the environment is already stocked with organisms. Working in the other direction, the higher the standing diversity is, the greater the competition for resources, and hence selection pressure. At these times there are also more, and more varied, genomes from which innovation may come.

There are times in the past when diversity seems to have increased slowly, and times when diversification has been rapid.

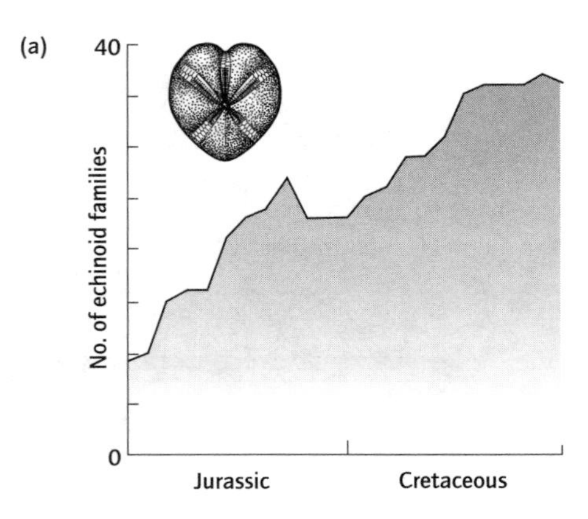

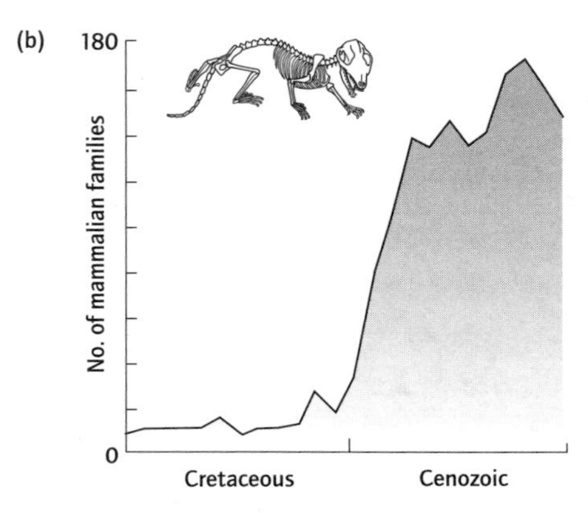

Fig. 16.4 Graphs of diversity through time for (a) Jurassic and Cretaceous echinoids, and (b) Cretaceous and Cenozoic vertebrates. Echinoid diversity increased slowly as a result of biological innovations related to burrowing. Mammalian diversity increased dramatically as the group radiated to fill empty ecospace left by the extinction of the dinosaurs.

Slow increases, such as those seen during the Jurassic and Cretaceous, seem to be a function of biological innovation resulting in niche diversification. As organisms evolve novel solutions to environmental problems, they increase potential and actual biodiversity. A good example of this is the evolution of a burrowing habit in echinoids, which led to an overall increase in diversity (Fig. 16.4a).

Times of rapid diversification usually appear to follow mass extinctions. In these cases, ecospace has been emptied by the huge, nonselective cull that has just been applied to the system and the survivors evolve quickly to fill the niche space. In the 10 million years after the dinosaurs went extinct, over 20 new families of mammals evolved, filling all the main empty niches from tree-dwellers to marine predators (Fig. 16.4b).

Mass extinctions

Extinction happens all the time, but there are rare examples of short periods of time where extinction rates have been very high. The definition of the rate and duration of these events is subjective, but most researchers agree that five mass extinction events have occurred since the Cambrian. These large faunal turnovers have been used by stratigraphers to define the boundaries of intervals of geological time. They occur at what we define as the end of the following geological periods: Ordovician, Devonian, Permian, Triassic, and Cretaceous (Table 16.1).

These mass extinctions are unique in detail but share the following characteristics:

1 A huge proportion of species became extinct – between 30 and 95%.

2 The extinctions seem to have operated across a wide range of environments and lifestyles.

3 The extinction event occurred rapidly and was probably caused by one or more physical factors.

One of the major problems with mass extinctions is defining cause and effect. By their nature, they are periods of time when conditions are unlike any modern analog. It is not known how ecosystems can be made to collapse in a catastrophic way, and it is unclear whether local disasters, such as a volcanic eruption, can be scaled up to help us understand global disasters. Intensive study of the boundary sections that represent mass extinctions rarely yield unequivocal data. Although extinctions have come to be associated with particular events, for example the assumed relationship between the end-Cretaceous event and a large meteorite impact, it is never possible to remove all uncertainty about whether the likely event actually caused the extinctions.

Table 16.1 A brief description of the five major mass extinction events of the Phanerozoic.

Mass extinction	Biological expression	Probable causes
End Ordovician	About 70% of marine species became extinct. Tropical faunas in general were badly damaged, especially reefs. Main affected groups were trilobites, graptolites, echinoderms, and brachiopods	A sudden, major glaciation, spreading from the South Pole. Most of the tropical belt disappeared. Sea levels fell, reducing the shallow shelf area, and cold water faunas moved to low latitudes, excluding warmer water ecosystems
End Devonian	A series of events lasting about 10 My. This extinction was characterized by low rates of origination, and extinction rates as such remained unexceptional. Cephalopods, fish, and corals were most affected	The least understood mass extinction. Sea bed anoxia or extraterrestrial impact have been cited. Another suggestion is that the marine ecosystem was badly affected by the rise of land plants and a short-lived drawdown of carbon dioxide
End Permian	The largest mass extinction, removing 95% of marine species and over 50% of marine families. Trilobites, cephalopods, corals, bryozoans, and crinoids were badly affected. Major faunal and floral overturn on land. Marks the boundary between dominance by the Palaeozoic and Modern fauna	Event may relate in some way to the supercontinent of Pangea, which would have affected world climates and oceanic conditions. The largest terrestrial igneous province was emplaced in Siberia at this time, and would have changed the climate over a range of timescales. There is controversial evidence for meteorite impacts at this boundary
End Triassic	A multiple event again. Most important on land, where floral overturn exceeded 95%. Around 30% of marine species became extinct, mainly reef-dwellers including ceratite ammonites, brachiopods, and bivalves	Recent work suggests that widespread submarine volcanic activity may have been responsible for massively elevated carbon dioxide levels at this time. This volcanic activity was a function of the breakup of Pangea
End Cretaceous	Dinosaurs, marine and aerial reptiles, ammonites, and belemnites became extinct. Brachiopods, bivalves, and foraminifera were severely affected	Most commonly attributed to a major meteorite impact into the Yucatan area of Mexico. Underlying sulfates and limestones would have vapourized to produce acid rain. Short-term cooling (from dust and sulfur dioxide) and long-term warming (from carbon dioxide) resulted. Also linked to the Deccan Igneous province in India

Life on land

The most significant event in the evolution of life during the Phanerozoic has been the colonization of land. Bacteria probably invaded fresh water and damp areas early in the Precambrian, and lichens were probably contributing to soil formation by 1,000 million years ago. However, larger animals and plants all migrated from the sea onto land during the Palaeozoic. In doing so, they radically increased the living space available on the planet. The innovations evolved to deal with this hostile environment altered the physical characteristics of the Earth, changing weathering cycles, the composition of the atmosphere, and climate.

There is evidence that plants began to migrate onto land during the Ordovician. By Silurian times, a widespread, low-lying plant cover was established wherever conditions were wet and by the Devonian plants had evolved into drier and higher locations and were beginning to form the first forests.

Animals followed plants into this new environment through the Devonian and Carboniferous. Millepedes, spiders, and other primitive arthropods were amongst the first to make the transition, and insects evolved on land during the Carboniferous. Worms, slugs, and snails all colonized the leaf litter of the early forests. Following these prey animals, came predatory tetrapods. In the Silurian, lobe-finned fish were probably capable of limited movement on land, and may have used this facility to escape marine predators or drying pools. The first amphibians are found in Devonian rocks. These were mainly fish-eaters who spent most of their time in the water, but their relatives soon began to exploit the resources available on the forest floor. The earliest reptiles are Carboniferous in age, and their terrestrial lifestyle was underscored by their ability to lay eggs on dry land.

A variety of routes onto land were used by organisms. Plants and vertebrates are thought to have migrated from fresh water, while terrestrial molluscs and arthropods have evolved from marine ancestors. Worms and other soft-bodied animals may have made the transition underground, moving through soft sediments into soils.

Physical challenges of life on land

Plants and animals had to overcome a similar set of physical challenges as they moved from life in water to life in air. These included the problems of support, of keeping themselves hydrated, of exchanging gases with the atmosphere, and of reproducing (Table 16.2).

Organisms need much stronger supports to stay upright in air than they do in the supporting medium of water. Plants developed tissues such as lignin and cutin, while animals strengthened their preadapted hard cuticles, in the case of arthropods, or internal skeletons, in the case of vertebrates.

Maintaining an internal watery environment is a challenge for all land-dwellers. Some have only partially solved this problem, and live in areas that are wet or that at least protect them from drying out, such as soils and leaf litter. Other small organisms have become adapted to withstand desiccation. Mosses are extremely tolerant of drying out, and will regenerate when water returns to their habitat. A variety of small mites are able to dry out completely, and will reanimate when immersed in water. Other forms produce eggs that can survive long periods of drought even though the adults die.

Most large land-dwellers have developed a waterproof membrane to keep their water inside. Plants have a waxy cuticle on their leaves and animals have waterproof cuticles or skin that lessen the risk or rate of desiccation. Animals have also evolved modified metabolisms that lose less water during digestion than is the case in aquatic organisms.

A problem that arises with this waterproofing is the difficulty of exchanging gases with the atmosphere. Plants need to extract carbon dioxide from the air and expel oxygen during photosynthesis. During respiration both plants and animals need to extract oxygen from the air and expel carbon dioxide. Most plants achieve this gaseous exchange through stomata, small holes in the waxy cuticle that are opened and closed by a pair of guard cells. Small animals such as spiders can rely on diffusion along a network of tubes within the body, but larger

Table 16.2 Summary of the main ways in which animals and plants have adapted to the challenges of life on land.

Challenge	Plants	Animals
Physical support	Woody tissues, lignin, and cutin (which may have evolved from precursor molecules involved in protection from UV radiation)	Arthropod hard cuticle and tetrapod skeletal strengthening. To some degree animals such as these are preadapted to live on land
Retention of water	Development of waxy cuticles, and behavioral methods of retaining water	Waterproof cuticle, water-retaining skin, behavioral methods, and modified excretion
Gas exchange	Via stomata in advanced plants	By diffusion if small, or via specialized organs such as lungs
Reproduction	Spores, protective seed pods, and internal fertilization	Waterproof eggs and internal fertilization

animals exchange gases with the atmosphere using specialized organs such as the vertebrate lung.

Reproduction in water can rely on this medium for transportation and for some degree of physical protection of the gametes. In air, these need to be protected by, for example, the application of protective coats such as spores for plants and amniotic eggs for some animals. In both plants and animals an innovative solution to the problems of terrestrial reproduction has been internal fertilization, where the adaptations of the adult protect the offspring.

Effects of life on land

The global effects of life on land have been progressive and dramatic, changing fastest during the Carboniferous as the major forest systems developed.

Widespread plant cover changes the nature of weathering patterns and of soils. Landscapes are stabilized by plants, and soils become more acidic and retain more water than before. As a result, rates of chemical weathering are increased and there is an increased sediment yield to rivers and the sea. The sediment produced is more mature than from unvegetated landscapes, with more of the chemically unstable minerals such as feldspars carried in a dissolved state and precipitated as clays. Increased weathering of silicate rocks adds calcium to the oceans and this is then bound into limestone. In this process, carbon dioxide is removed from the atmosphere.

Carbon dioxide is also removed by the plants themselves and held within the biomass of the forest. It can then be buried in the form of organic-rich deposits such as coal or gas (oil is formed by plankton). Two effects of this are to cool the planet by removing a greenhouse gas from the atmosphere, and to produce large reserves of fossil fuels.

While some of the organic carbon produced by forests is retained and buried there, much of it is removed by erosion and is transported by rivers to the sea. This in turn stimulates productivity in the oceans and increases demand for oxygen. Oceans become more prone to run out of oxygen close to these nutrient-rich sites, a process known as eutrophication. This can locally kill large numbers of marine organisms and can lead to black shales being deposited in the rock record, some of which are oil-producing when buried. This increased productivity in the oceans extracts even more carbon dioxide from the air and tends to cool the planet even more (Fig. 16.5).

During the Devonian, when runoff first began to carry large volumes of organic matter into the oceans, there was a prolonged evolutionary crisis amongst the plankton. The two events may be linked. More controversially, some researchers believe that the removal of carbon dioxide by plants during the Carboniferous was intense enough to have caused the major Permo-Carboniferous ice age. Research seems to show that carbon dioxide levels in the atmosphere fell very quickly at this time, though it is not possible to know whether this contributed to cooling the planet or was the main cause of the glaciation.

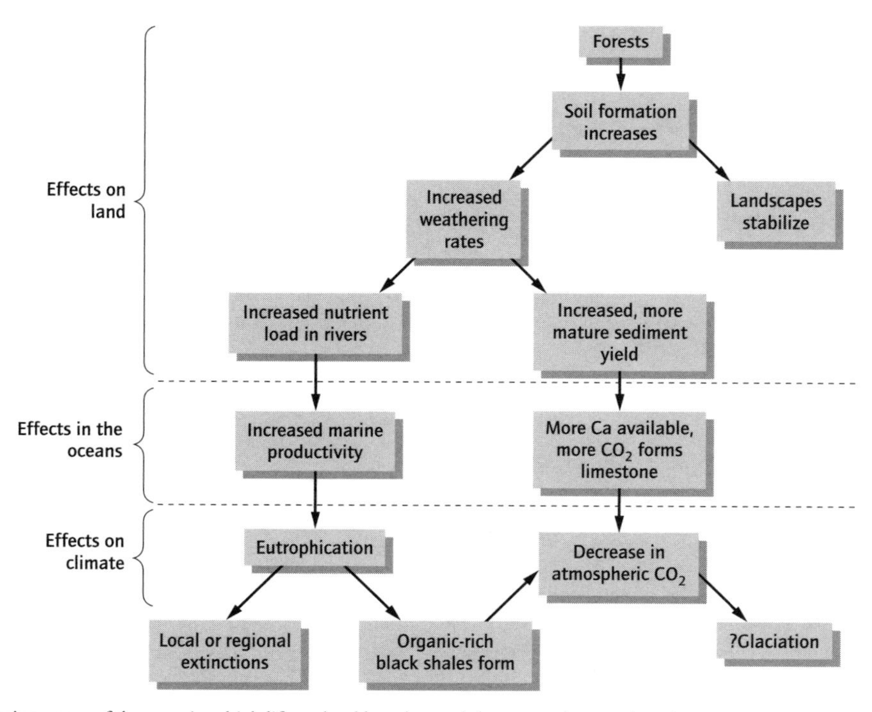

Fig. 16.5 Flow chart showing some of the ways in which life on land has changed the atmosphere and environment.

Glossary

Abiotic – nonliving.
Adaptive radiation – evolution of numerous species from a common ancestor.
Biodiversity – diversity of living organisms.
Biotas – assemblages of organisms.
Biotic – from living organims.
Darwinian selection – selection of organisms based on the relative contribution of an organism to the gene pool of the next generation.
Ecospace – area available for organisms to occupy. This is partly a physical space, and partly a series of interactions between organisms.
Evolution – processes of change in living organisms over time.
Niche – also called ecological niche. The sum of an organism's use of all available environmental resources.
Species – organisms with similar anatomical characteristics that can, potentially, interbreed.
Species diversity – number and relative abundance of species in a biological community.
Taxonomy – science of naming and classifying organisms.

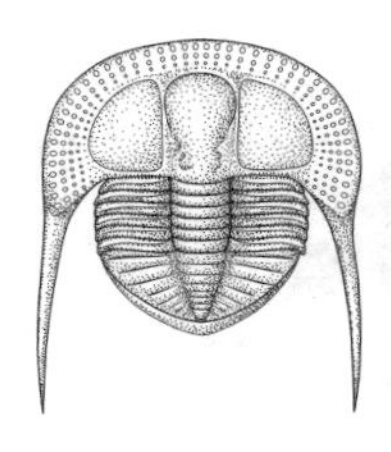

Reading list

Ausich, W.I. and Lane, N.G. (1999) *Life of the Past.* Prentice Hall, Englewood Cliffs, NJ.

Benton, M. and Harper, D. (1997) *Basic Palaeontology.* Addison Wesley, Longman.

Brenchley, P.J. and Harper, D.A.T. (1998) *Palaeoecology: Ecosystems, Environments and Evolution.* Chapman and Hall, London.

Briggs, D.E.G. and Crowther, P.R. (1990) *Palaeobiology I.* Blackwell Science, Oxford.

Briggs, D.E.G. and Crowther, P.R. (2001) *Palaeobiology II.* Blackwell Science, Oxford.

Clarkson, E.N.K. (1998) *Invertebrate Palaeontology and Evolution.* Blackwell Science, Oxford.

Cowen, R. (2000) *History of Life,* 3rd edn. Blackwell Science, Oxford.

Doyle, P. (1996) *Understanding Fossils. An Introduction to Invertebrate Palaeontology.* Wiley, Chichester, UK.

Gee, H. (2000) *Shaking the Tree, Readings from Nature in the History of Life.* Chicago University Press, Chicago.

McKinney, F.K. (1991) *Exercises in Invertebrate Paleontology.* Blackwell Scientific Publications, Oxford.

Prothero, D.R. (1998) *Bringing Fossils to Life: an Introduction to Paleobiology.* W.C.B./McGraw-Hill, New York.

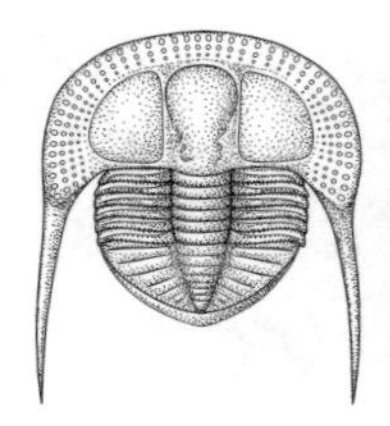

Geological timescale

Eon	Era	Period	Epoch	
				Present day
Phanerozoic	Cenozoic	Quaternary	Holocene	0.01 Ma
		Tertiary	Pleistocene	1.8 Ma
			Pliocene	5.3 Ma
			Miocene	23.8 Ma
			Oligocene	33.7 Ma
			Bocene	54.8 Ma
			Palaeocene	65 Ma
	Mesozoic	Cretaceous		144 Ma
		Jurassic		206 Ma
		Triassic		248 Ma
	Palaeozoic	Permian		290 Ma
		Carboniferous		354 Ma
		Devonian		417 Ma
		Silurian		443 Ma
		Ordovician		490 Ma
		Cambrian		543 Ma
Proterozoic		2,500		
Archean		3,800		
Hadean		4,600		

Proterozoic, Archean and Hadean: Also known collectively as the Precambrian

Ages based on dating techniques and fossil records as of 1999. Ma, + millions of years.

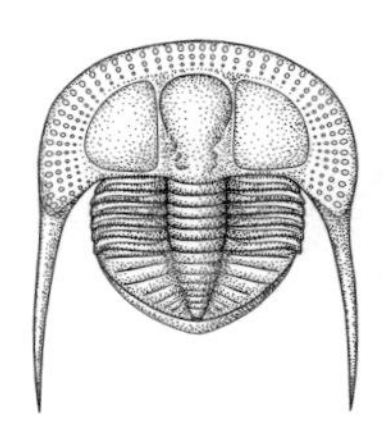

Index

Note: Page numbers in *italic* refer to figures and those in **bold** refer to tables

abiotic effects 140, 146
acanthodians 88, 89, 100
Acanthostega 90
acheulian tools 100
acritarchs *113*, 114, 133
actinopterygians 89, 100
adaptive radiation 4, 10, 15, 16, 146
adductor muscles
  bivalve *64*, 67, 78
  brachiopod 36, 42
  ostracodes 120
agnathans 88, 100
agnostids 57
Agnostus *55*, *56*, 58
Agrichnia **128**
algae 115, 116
allopatric speciation 14, 16
alternation of generations
  cnidarians *21*
  land plants 104
alveolus *73*, 78
Amaltheus 76
amber 8, 9, 10
ambulacra 44, 48, 52
ammonites 62, 70, 76, 77
  heteromorph 71, *72*
  life attitudes *72*
  life habits *72*
  morphology 71
  oxycone *71*
  palaeoecology 72
  platycone *71*
  serpenticone *71*, *72*
  sexual dimorphism 71
  sphaericone *71*, *72*
Ammonoidea **62**, 68, 70
  evolutionary history 68, 70
  heteromorph 78
  sutures 70
amniotes 92–4
amphibians 90, 91, 144
Amphoracrinus 50
Ampyx 56
anal tube 52
anapsids *91*, 92, 93, 100
ancestrula 31, 32, 34
angiosperms 102, **103**, 104, 108, 109, 112, 123
animals, classification 134
Annularia 110
antennae, trilobite 55
antennule 126
anthocerophytes **102**, 105
Anthozoa 21, 22
anthracosaurs 90
apertures 118, 126
apical system 52
Apiocrinites 50
aragonite 7, 17, 22, 24, 26, 27, 31, 65
Arca **65**
Archaefrutus *109*
Archaeocidaris 51
archaeocyathids 17, 19, *138–9*
archaeocytes 18
archaeogastropods 63, 76
Archaeopteris **102**
Archaeopteryx 9, 99
archosaurs 98
Arenicolites *129*
Aristotle's lantern 48, 52
arms
  brittle star 47
  crinoid *46*
Arthropoda 53
articular bone 95
Articulata 35
ascon-grade organization, sponges 18, 20
Asteriacites **128**, *129*
Asteroceras obtusum 12
Asteroidea **44**, 45, 47
atoll 26
australopithecines 96
Australopithecus afarensis 96, **97**
autothecae *80*
autotroph 126
avicularia 31, 32, 34
axial lobe, trilobite 54, *55*, 60

backbone 87
banded iron formations 131
basal plates 52
basking shark 89
Bathropyramis 124
beak, brachiopod 37, 42
behavior, information loss concerning 5
belemnites 68, 73, *77*
bennettitales **103**, 108, *109*
benthic organisms 10, 17
benthos 126
Bergamia *56*
Beyrichia 125
biases in the fossil record 7
bilateralia 136
biodiversity 146
biogeography 5
biostratigraphy 3, 4, 10, 57, 70, 113, 123
biota 146
biotic 146
biozones 4, 10
bird-hipped dinosaurs 98, 99, 100
birds 99
  flightless 95
bithecae *80*
Bitter Springs Chert 105
bivalves 35, 39, 62
  boring 67
  dentition **65**
  ecology 66
  epifaunal 67
  evolutionary history 65
  infaunal 66
  morphology 64
  rudist 65, 74
  shells 62, 65

bivalves (*cont'd*)
  swimming *67*
  trace fossils **128**
black shales 114, 145
Blastoidea **45**
body chamber
  ammonite *71*
  Nautilus *69*
body fossils 2, 127
Bohemograptus 85
Bolivina 124
bony fish 88, 89
brachial groove 36
brachials 52
brachidium 42
brachiopods 3, 31, 35
  Atlantic *5*
  classification 35
  community palaeoecology 39
  ecology and palaeoecology 38
  epifauna 38
  evolution 39
  external morphology *37*
  internal morphology 36
  lifestyle and morphology **38**
  Pacific *5*
brittle stars **44**, 45, 47
brood pouch, ostracodes 120
bryophytes **102**, 105
bryozoans 31
  ecology 32
  as environmental indicators 33
  evolution 33
  morphology 32
Buccinum *75*
Burgess Shale 9
burrows 127
byssal notch (gap) 67, 78
byssus threads 67
Bythocertina 125

caecilians 90
Calamites 107, 110
Calcarea *17*, 20
calcite 7, 17, 18, 22, 23, 24, 31, 54, 65, 115
calcite compensation depth 117, 126
calcium 145
calcium carbonate 27
calcium phosphate 87
calice *24*, 30
Callistophyton *108*
Calymene *55*, *56*, 59
calyx *46*, 52
camaroids 79
cambium 104, 112
Cambrian
  diversity 140
  explosion 138–9
  fauna 137, 141
  palaeogeography *138–9*
  trilobites *57*
camera *69*, 78
carapace 93, 120, 126
carbon dioxide 144, 145
carbon isotopes **132**
Carboniferous
  coal forests 107
  diversity *140*
  land animals 144
Carrier's constraint 92, 95, 98
cartilaginous fish 88, 89
cateniform colony 30
Caulostrepsis *129*
CCD 117, 126
Cenozoic, diversity *140*
center of bouyancy 72
center of gravity 72
centrales 116, 126
cephalic shield 81, 86
Cephalodiscus 81, **82–3**
cephalon 54, *55*, 56, 60
cephalopods 62
  evolutionary history 68
  morphology 68
ceratitids 70
cerioid colony 30
chamber 126
Cheilostomata 31, 33
chemical fossils 2, 133
cherts, nodular 117
chlorophyll 2, 10
chloroplasts 112, 133
choanocytes 18
chondrichthyans 89, 100
Chondrites **128**
Chonosteges **38**
chordates 87
chronostratigraphy 3, 10
Cidaroidea 51
citric acid cycle 133, 136
clades 92
cladistics 11, 12, 16
cladogram 12, 16
classes 12
classification 11, 12, 146
clays 145
Climacograptus **82–3**, 85
climate and corals 27
Clonograptus **82–3**
Clorinda 39
club mosses 102, 107, 110, 111, 112
Clypeus 51
cnidarians 21
coal 145
coal forests, Carboniferous 107
coccolithophores *113*, 115, 125
coccoliths 115
coccosphere 115
codons 16
coelocanth 89, *91*
coenenchyme 23, 24, 30
coenosteum 26, 30
Colaptomena 41
Coleoidea **62**, 68, *73*
collagen 79, 80, 86
columella *24*, 30
columnals 52
commissure 35, 37, 42
common ancestor 132
community structure, information loss concerning 5
compression fossils 101
concentration deposits *see* Konservat-Lagerstätten
cones **103**
Confusastraea *26*
coniferophytes 112
conifers 102, **103**, 108, 109
conodont animals 87, 88, 122
conodonts 88, 113, 122
  coniform elements 122, 126
  pectiniform elements 122
  ramiform elements 122, 126
conservation traps 8
conservative gene sequences 13
Cooksonia **102**, 105, 106
coprolites 2, 127
corallite 22, 23, 24, 25, 26, 30
corallum 22, *23*, *24*, *26*, 30
corals 21
  ahermatypic 22, 30
  as climatic indicators 27
  colonial 19
  hermatypic 22, 30
  morphology and evolution 22
  provincialism 26
Cordaites 107, 108
cortical bandages 86
Coscinodiscus *116*, 124
Cosmorhaphe *129*
costa *64*
Costricklandia *39*
Crania **38**
Craniiformea 35, *37*, 39
Crassostrea *67*
creodontids 95, 100
Cretaceous, diversity *140*, 142
crinoids 45
  dicyclic *46*
  evolution 45, 46
  feeding 46
  monocyclic *46*
  morphology 46
  water vascular system and lifestyle **44**
Crioceras *72*
crocodiles 99
Crossopodia *129*
crustoids 79
Cruziana **128**
  ichnofacies *129*
Ctenostomata 33
Cubichnia **128**
cuticle 54, 104, 105, 112, 144

cutin 144
cuttlefish 68, 73
cyanobacteria 5, 10
Cybeloides 56
Cycadeoidea **103**, *109*
cycadeoids 108
cycads 102, **103**, 108
Cyclomedusa 135
Cyclopyge 59
Cyclostomata 33, 34
cynodonts 94, 95, 100
Cypridina 125
Cypris 125
cyrtograptids **82–3**
Cyrtograptus **82–3**, 85
cystoids 79
cysts, dinoflagellate 115
cytoskeleton 133

Dactylioceras 72, 76
Dalmanites 58
Darwin, Charles 13
Darwinian selection 13, 16, 146
Deinonychus 99
Deiphon 59
delthyrium 37, 42
deltidial plates 37, 42
deltidium 37, 42
Demospongia **17**, 18, 20
dendroids 79, *80*, **82–3**, 84
dentary 95, 100
dentine 88
deposit-feeding, trilobites 56
desiccation 144
Devonian
  diversity *140*
  land animals 144
  land plants 144
  plankton 145
  reefs 25
  trilobites 57
diagenesis 6, 10
diapsids *91*, 92, **93**, 98–9, 100
diatomites 116, 126
diatoms 19, *113*, 116, 124
Dibunophyllum 29
Dicellograptus **82–3**
dichotomous branching 112
Dickinsonia 135
Dicranograptus **82–3**, 84
Dictyodora **128**
Dictyonema 84
diductor muscles 36, 42
Didymograptus **82–3**, 84
Dimetrodon 94
dinoflagellates *113*, 114, 115, 124, 133
dinosaurs 95, 98, 99
dinosterane 114
Diplichnites **128**
Diplocraterion **128**, *129*
diplograptids **82–3**
diploid cells 112
Diptera 9
dissepiments 23, 24, 26, 30, 80
diversity 11, 140, 142
DNA 13, 16, 133, 136
Domichnia **128**
double fertilization 112
Dunkleostus 89

Earth, formation 131, **132**
ecdysis 53, 60
echinoderms 24, 43
  characteristics **45**
  classification 45
  evolutionary history 45
  symmetry 43, 44
echinoids 45
  deep infaunal *49*, **49**
  dicyclic 52
  ecology 49
  epifaunal *49*, **49**
  evolution 45, 49, 142
  irregular 52
  monocyclic 52
  morphology 48
  regular 52
  shallow infaunal *49*, **49**
  water vascular system and lifestyle **44**
ecospace 146
Ediacaran fauna 131, 134, 135
eggs
  amniotic 92
  in fossil record 92
  ostracodes 120
eleutherozoans 45
Elphidium 124
Emiliana 125
enamel 88
endoskeleton 43, 45
endosymbiosis 133
endothermy 98
Ensis 66, 74
Entobia 129
enzyme 136
Eocene, climate change 95
Eocoelia 39
Ephedra 109
epifauna 38, 126
epirelief 127, 130
Equilibrichnia **128**
Equisetum 110
equivalve 78
eukaryotes 131, *132*, 133, 134, 136
eutrophication 145
evolution 11, 13–15, 142, 146
evolutionary conservatism 57
evolutionary palaeoecology 4
evolutionary relationships 4
exoskeleton 10, 53, 54
eyes, trilobite 54, *55*, 56
fasciculate colony 30
fasciole 52
Favosites 23, 28
fecal pellets 134
feldspars 145
Fenestella 34
Fenestrata 33, 34
ferns 102, **103**, 104, 107, 111
filter-feeding, trilobites 56
fins 89, *90*
fish 88–9
fixed brachials *46*
fixed cheek (fixigena), trilobite 54, 60
flagella 18, 20
flint 19
flowers 104
Fodinichnia **128**
fold 37, 38, 42
foot
  bivalve 64, 65, 66
  gastropod 63
  mollusc 61
foramen 42
foraminifera *113*, 118–19, 124
forests 104
form genus 85
formations 3
fossil fuels 145
fossil Lagerstätten 8
fossilization process *6*
fossils
  definition 1
  information from 4
  preservation 6
  types 2
fossulae *24*, 25
free cheek (librigena), trilobite 54, 60
frog 90
frustule 116, 124, 126
Fugichnia **128**
full relief 130
fusellar increments 86

gametes 104, 112
gametophyte 104, 109, 112
gas 145
gas exchange 144
Gastrioceras 76
gastropods 62
  classification 63
  ecology 63
  evolutionary history 63
  morphology 63
gene sequences 136
genera 11, 12
genes 13, 15
genetic code 136
genital plates 52
genome 13, 16
geological time 3, 4
Gigantoproductus 40

gills
  bivalve 64
  Nautilus *69*
ginkgoes 102, **103**, 108
glabella 54, *55*, 60
Globigerina 124
gnathostomes 88–9, 100
gnetales **103**, 109
golden spike 3, 10
goniatitids 70
government rock 40
graptolites 3, 79, 84–5
  biserial 86
  evolutionary events **82–3**
  mode of life 81
  morphology 80
graptoloids 79, *80*, 81, **82–3**
Great Barrier Reef 26
green algae 105
growth lines
  ammonite *71*
  bivalve *64*
Grypania 131
Gryphaea *67*
Gymnolaemata 31, 32, 33
gymnosperms 102, **103**, 104, 108–9, 112, 123

habitat, information loss concerning 5
hagfish 88
Halysites *23*, 28
Hamulina *71*
haploid cells 112
heart urchins 48
Heliolites 23
helioplacoids 45
Hemichordata 79
Hemicidaris 51
Heminthoida *129*
hepatophytes **102**, 105
Hertzina 122
heterococcoliths 126
heterotroph 126
Hiatella *67*
Hildoceras *77*
hinge
  astrophic 42
  brachiopod 37
  strophic 42
holdfast
  corals 24, 26
  crinoid *46*
  graptolite 80, 86
  sponges 18
holococcoliths 126
Holothuroidea **44**, 45
hominids 96, **97**, 100
Homo erectus 96, **97**
Homo habilis 96, 97
Homo sapiens 96
Homo sapiens neanderthalensis **97**
Homo sapiens sapiens **97**

hood *69*, 78
Horneophyton 106
hornworts **102**, 105
horsetails 102, **103**, 107, 110
humans 96
Hygromia 75
Hylonomus 91, 92
Hyphantoceras *71*
hyponome 68, 69, *73*, 78
hyporelief 127, 130
hypostome 54, *55*, 56, 60
Hystrichosphaeridium 124

Iapetus Ocean **82–3**
ice age, Permo-Carboniferous 145
ichnofacies, marine 129
ichnofossils *see* trace fossils
ichnogenera 127, **128**, 130
ichnospecies 127, 130
ichthyosaurs 99
Ichthyostega 90, *91*
iguanodonts 99
ilium 100
impression fossils 101
Inarticulata 35
inequivalve 78
infaunal organism 10, 126
information loss during fossilization 5
insects 144
interambulacra 48, 52
interarea 37, 42
interbrachials *46*
Isastraea 28
ischium 100
Isograptus **82–3**
isotopes 27, **132**, 136
Isua Supracrustals **132**
iterative evolution 31

jaws 88, 89
  articular 100
  mammal 95
  reptile *95*
  squamosal 100
jellyfish 21, 135
Jurassic, diversity *140*, 142

Kilbuchophyllia 22
kingdoms 12
Kochiproductus **38**
Kodonophyllum *27*
Konservat-Lagerstätten 8, 9, 10
Konzentrat-Lagerstätten 8, 10
Kosmoceras *77*
Kreb's cycle 133, 136

lamprey 88
land
  colonization 144
  global effects of life on 145
  physical challenges of life 144

land plants 101
  classification 102, **102–3**
  colonization of the land 104, 105
  earliest 105
  flowering **103**
  key steps in evolution 104
  life histories 104
  nonvascular 102, 105, 112
  spore-bearing 107
  vascular 102, **103**, 105, 106
lappets 71, 78
larvae 21
leaf cushions 111
leaves 105
  evolution 104
  lycopod 107
Lepidodendron *101*, 107, 110, 111
leucon-grade organization, sponges 18, 20
life
  evidence for early life 132
  evidence for timing of evolution **132**
  origin of complexity 133
lignin 104, 105, 112, 144
limb bones, vertebrate 90
limestone 145
Lingula 38, 39, 40
Linguliformea 35, 37, 39
lissamphibia 90
lithostratigraphy 3, 10
Lithostrotion 29
liverworts **102**, 105
living fossils 38, 40, 89
lizard-hipped dinosaurs 98, 99, 100
lobe-finned fish 88, 89, 90, 144
lobes 70
lophophore 31, 36, 37, 42, 81, 86
Lorenzinia *129*
low energy environments *7*
Lower Silurian, brachiopod palaeocommunities 39
Ludwigia *72*
lungfish 89, 90, *91*
Lutraria **65**
lycophytes (lycopods) 102, 107, 110, 111, 112
Lytocrioceras *72*

macroconchs 71, 78
macroevolution 11, 15
Macroscaphites *71*, *72*
macrozooplankton 79
maculae 32
madreporite 44, 52
Magellania **38**, 40
mammals 95, 142
mantle 42, 61, *64*, 73, 78
mantle cavity 61, *73*, 78
  cephalopod 68
  gastropod 63
  Nautilus *69*
marine organisms *7*
Mariopteris 111

marsupials 95, 100
Marsupites 50
mass extinction events 137, 142, 143
  definition 16
  end-Cretaceous 95, 99, 140, **143**
  end-Devonian 140, **143**
  end-Ordovician 57, 140, **143**
  end-Permian 35, 39, 57, 140, **143**
  end-Triassic **143**
  and fossil record *4*
  graptolites **82–3**
masseter 95, 100
medial growth, corals 23
Medullosa 108
medusoid *21, 22*
megaspores 123, 126
meiosis 112
meristem 112
mesogastropods 63, 75
Mesozoic 98
  community structures 4
metazoans 134, 136, 138
Micraster 51
microconchs 71, 78
microevolution 11, 14
microfossils 113–26
microinvertebrates *113*, 120–1
micropalaeontology 113
microvertebrates *113*, 122
millerettids 93
mitochondria 133, 136
Modern fauna 137, 141
molecular clock 13, 16, 136, 138
molluscs 61–78
  basic morphology 61
  classification 62
  origins 61
  shell growth 61, 62
molting 53, 60
monograptids 81, **82–3**
Monograptus **82–3**
monopodial branching 112
monotremes 95, 100
Montlivaltia 29
morphology, information loss concerning 5
moss animals *see* bryozoans
mosses **102**, 105
mousterian tools 100
multicellular organisms 134
mural pores 30
muscles
  bivalve *64*, 67, 78
  brachiopod 36, 42
  diductor 36, 42
  ostracodes 120
mutation 13, 16
Mya 66, 74
Mytilus **65**, *67*

nannofossils 126
nannoliths 115
nannoplankton 126
Nassellar 117, 126
natural selection 13, 16, 146
Nautiloidea **62**, 68, 69
Nautilus 69
nema 80, 81, 86
neogastropods 63
Neohibolites 77
Neotrigonia **65**
Nereites **128**
  ichnofacies *129*
niche 146
nodular cherts 117
nomenclature 12
Normalograptus **82–3**
notochord 87
notothyrium 37, 42
nucleus 133

obrution deposits 8
occular plate 52
octopus 73
oil 145
oldowan tools 100
Olenus 56, *56*
oozes
  diatom 116
  radiolarian 117
operculum 32, 34, *63*, 78
Ophiomorpha 129
Ophiuroidea **44**, 45, 47
Opipeuter 56
opisthobranchs 63, 75
orders 11, 12
Ordovician
  colonization of land 144
  diversity 140, 141
  trilobites 57
organelles 112, 133, 136
ornithischians 98, 99, 100
osculum 18, 20
ossicles 43, 45, 52
osteichthyans 89, 100
ostia 18, 20
Ostlingoceras 71
ostracodes 113, 120–1, 125
outgroup 12
oxygen 131, 133, 144, 145
Ozarkodina 122

palaeobathymetry 119, 126
palaeoclimatology 27
palaeocommunities 39
Palaeodictyon **128**, *129*, 130
palaeoecology 4
palaeoenvironment 5
palaeontology 132
Palaeosmilia 28
Palaeozoic
  colonization of land 144
  community structures 4
  diversity *140*
  fauna 137, 141
pallial line *64*, 78
pallial sinus *64*, 66, 78
palynology 123, 126
Pangea 95, 140
Paradoxides 58
paragaster 20
parallel evolution 95
Paranthropus 96, **97**
pareiasaurs 93
Pascichnia **128**
Patella 76
Pecten **65**, 67
pectiniform 126
pedicellariae 48, 52
pedicle 36, 38, 42
pedicle foramen 37
pelagic organisms 3, 10, 126
pelmatozoans 45
pelycosaurs 94, 100
pennates 116, 126
Pentacrinites 50
pentadactyl limb 90
Pentamerus 39, 40
pentaradiate symmetry 43, 44
perimineralization 101
peripheral growth, corals 23
periproct *48*, 52
Perisphinctes 71
peristome *48*, 52
Permian, diversity 140
petrification 101
Phanerozoic
  diversity 140–1
  life 137–46
Phillipsia 59
phloem 104, 112
Pholas 67
Pholiderpeton 90
phosphate *138–9*
photic zone 116, 126
photosynthesis 10, 104, 131, 132, 133, 136, 144
phragmocone 71, 73, 78
Phycodes 129
Phycosiphon **128**
phyla 12
phyletic gradualism 14, 16
Phyllograptus **82–3**
phylogeny 12, 16
phytoplankton 114, 126
Pikaia 87
pinnule 46, 52
Pinus succinifera 9
pipe rock *129*
pith cast 110
placental mammals 95, 100
placoderms 88, 89, 100
planktonic organisms 126
Planorbis 75

plants
  colonization of land 144
  microfossils *113*, 123
  see also land plants
plastids 136
plastron *48*, 51, 52, 93
plate tectonics 10
plesiosaurs 99
pleural lobes, trilobite 54, *55*, 60
pliosaurs 99
pollen 105, 112, 113, 123
  analysis 123
  monoporate **123**
  saccate **123**
  triporate **123**
pollen diagrams 123
pollination 109
polychaetes, trace fossils **128**
Polygnathus 122
polyp 21, 22, 24, 30
Porifera 17
prebilateralia 136
Precambrian
  diversity *140*
  life 131–6
Precambrian–Cambrian boundary 130, 131, 138–9
predation marks 127
prepubic process 100
preservation potential 7
Pricyclopyge 56
primary producers 126
primates 96
primitive characters 12
procolophonids 93
Proetus 56
progymnosperms **102**, 107, 108
prokaryotes 131, 132, 133, 136
pro-ostracum 73, 78
Prorichthofenia 41
prosobranchs 63, 75
protists 113
  animal-like 117–19
  plant-like 114–16
protoconch 69, 71, 78
Protogyrinus 90
protoplasm 126
pseudopodia 118, 126
pteridophytes 102, **103**, 104, 107, 111
pteridosperms **103**, 108, 111
pterobranchs 79, 81, **82–3**
pterosaurs 98–9
pubis 100
pulmonates 63
punctae 37
punctuated equilibrium 14, 16
Pycnocrinus 46
pygidium 54, *55*, 60

quadrate bone 95, 100
Quercus 123, 125

radial plates 52
radiolaria 19, *113*, 117
Radiolites 74
radula 78
Rafinesquina **38**
Rastrites 85
ratites 99, 100
ray-finned fish 89
rays 89
red tides 115
reefs 19, 25, 26, *27*
Repichnia **128**
reproduction
  on land 145
  sexual 133
reptiles 92–4, 98–9, 144
  marine 99
reptilomorphs 90
Retiolites **82–3**
retiolitids **82–3**
Rhabdopleura 81, **82–3**
rhabdosome 80, 86
Rhaphidonema 20, **82–3**
Rhaphoneis 116
rhipidistians 90, 91, 100
Rhizocorallium 129
rhizoid **102**, 112
rhizome 106, 112
Rhynchonelliformea 35, 36, 37, 39
Rhynia 106
Rhynie Chert 106
rhyniophytes **102**, 106
RNA 13
Rodinia 140
roots 105
rostroconchs 65
rostrum (guard) 73, 78
Rugosa 21, 22, 24, 25, 27
Rusophycus **128**

Saccocoma 9
saddles 70
saetograptids **82–3**
Saetograptus **82–3**
salamander 90
sand dollar 48
sarcopterygians 89, 100
sauropods 98, 99, 100
sauropterygians 99
Sawdonia 106
Scaphites 77
Scleractinia 21, 22, 26, 27
sclerocytes 18
Scottognathus typicus 122
scute 93
sea anemones 21
sea cucumbers **44**, 45
sea lilies *see* crinoids
sea urchins *see* echinoids
secondary palate 100
sediments 3

seed ferns **103**, 108, 111
seeds 104, 105, 108, 109
segments, trilobite 55
Selenopeltis 55
semirelief 130
Sepkoski's curves 140
septa 22, 23, 24, 25, 26, 30, 78
setae
  Nautilus 69
  ostracodes 120, 126
sexual reproduction 133
shared derived characters 12
sharks 89
shells 7
  ammonite 71
  Ammonoidea 70
  bivalve 62, 65
  brachiopod 37
  Coleoidea 73
  gastropod 63
  mollusc 61, 62
  Nautiloidea 69
sicula 80, 81, 86
Sigillaria 107, 111
silica 7, 17
  biogenic 19
  oceanic areas of high extraction by plankton *116*
Silurian
  diversity 140
  land animals 144
  land plants 144
siphon 78
  bivalves *64*, 65
  gastropod 63
siphonal canal 78
Siphonia 20
siphuncle 69, 78
  ammonite 71
  Ammonoidea 70
  belemnite 73
  Nautilus 69
sister groups 12
skeletons 138
Skolithos **128**
  ichnofacies *129*
skull, tetrapod 92, **93**
soils, effects of plant life 145
Solnhofen Lithographic Limestone 9
speciation 14
species 11, 12, 16, 146
species diversity 146
Sphenophyllum 110
sphenophytes 102, **103**, 107, 110
Sphenopteris 111
spicules 17, 18
spines
  brachiopod 38
  echinoid 48
  trilobite 54, *55*, 58, 59
spiralia 41, 42

Spirifer 41
Spirograptus **82–3**
Spirorhaphe **128**, *129*
spondylium 40, 42
sponges 17
  cell types 18
  hexactinellid (glass) *17*, 18, 19
  morphology 18
  reef-building 19
  siliceous 19
spongocytes 18
sporangia **102**, **103**, 105, 107, 111, 112
spores **103**, 104, 105, 112, 113, 123
Sporogonites *105*
sporophyte 104, 112
Spriggina 135
Spumellar 117, 126
squid 68, 73
stable isotope analysis 119
stagnation deposits 8
starfish **44**, 45, 47
Staurograptus **82–3**
stegosaurs 99
stem, crinoid 46
Stenolaemata 31, 32, 33
stereom 43, 45, 52
Stigmaria 107, 110
stinging cells 21
stipe 80, 86
stomata 105, 106, 112, 144
Stomatopora 34
strata 4
stratigraphy 3
stromatolites 132
stromatoporoids 17, 19, 23, 25
sulcus 37, 38, 42
suspension-feeding 46
sutures 54, *55*, 78
  ammonite *71*
  Ammonoidea 68, 70
  trilobite 60
sycon-grade organization, sponges 18, 20
symbiotic algae 19, 22
sympatric speciation 14, 16
synapsids 91, 92, **93**, 94, 100

tabulae 22, 23, 24, 25, 27, 30
Tabulata 21, 22, 23
taphonomy 6, 10, 101, 115
Taphrhelminthopsis *129*
taxon (taxa) 10
taxonomy *see* classification
tegmen 52
teleost fish 89
temnospondyls 90
tentacles, *Nautilus* *69*
Teredo 74
test 43, 52, 126
  asteroid 47
  echinoid 48
  foraminiferan 118, 119
  radiolarian 117
Tethyan Ocean 96
Tetragraptus 84
tetrapods 88, 90
Tetrarhynchia 41
Thalassinoides **128**, *129*
thallus **102**, 112
thecae 52, 80, 81, **82–3**, 86
Thecosmilia 29
therapsids 94, 100
theropods 98, 99, 100
thorax, trilobite 54, *55*, 60
Thrinaxodon 94, 95
tissues resistant to decay *7*
tool-making 96
tooth and socket, bivalves *64*, 65
tortoises 93
trace fossils 2, 5, 126, *138–9*
  ethological (behavioral) classification 128
  evolution 130
  marine ichnofacies classification 129
  preservation 127
Trepostomata 33
Treptichnus pedum 130
Triassic 137
  diversity *140*
Tribrachidium 135
Triceratops 99
trilete 112
trilobites 3, 53, *138–9*
  Atlantic *5*
  body fossils 2
  diversity through time *57*
  evolution 57
  exoskeleton *7*
  mode of life 56
  morphology 54, *55*
  Pacific 5
  trace fossils 2
Trinucleus *55*, 58
Trypanites *129*
  ichnofacies *129*
tube feet 43, 44, 46, 47, 48, 52
tuboids *79*
Turritella 75
turtles 93, 99
type sections 3
Tyrannosaurus rex 99

umbilicus *71*
umbo 37, 38, 42, *64*, 78

valves
  brachiopod 36, 37, 38
  dorsal (brachial) 42
  ventral (pedicle) 42
Vargula 120
vascular system, land plants 105, 112
Vendobionta 135
venter 78
Venus **65**, *66*
vertebrates 87–100
virgella **82–3**
virgula 80, 86
viscera 52, 78

water vascular system 43, 44, 52
weathering 145
Wenlock Limestone 27
Wenlock Series 3
Westlothiana 92
whorl 78

xylem 104, 106, 112

yolk 92

zoarium 32, 34
zone fossils 3, 10, 126, 130
zooecium 32, 34
zooids 31, 32, 86
Zoophycos 130
  ichnofacies *129*
zooplankton 126
zooxanthellae 22, 26, 27, 30
zosterophylls **102**, 106
zygote 104, 112